Quick Review on

Industrial Pharmacy - I

Quick Review on

Industrial Pharmacy - I

Ajay Semalty

M. Pharm., MBA, Ph. D., PDF (Japan)

Course Coordinator- industrial Pharmacy-I (SWAYAM MOOC)
Assistant Professor,
Department of Pharmaceutical Sciences,
H.N.B. Garhwal University (A Central University)
Srinagar (Garhwal), Uttarakhand
Visiting Scientist
Faculty of Pharmacy, Meijo University Nagoya, Japan

PharmaMed Press

An imprint of Pharma Book Syndicate

A unit of BSP Books Pvt. Ltd.

4-4-309/316, Giriraj Lane,
Sultan Bazar, Hyderabad - 500 095.

Quick Review on **Industrial Pharmacy - I**
by Ajay Semalty

Published by

PharmaMed Press
An imprint of Pharma Book Syndicate
A unit of BSP Books Pvt. Ltd.
4-4-309/316, Giriraj Lane, Sultan Bazar, Hyderabad - 500 095.
Phone: 040-23445600, 23445688; Fax: 91+40-23445611
E-mail: info@pharmamedpress.com
www.pharmamedpress.com/pharmamedpress.net
www.pharmamedpress.com/pharmamedpress.net

ISBN: 978-93-91910-32-7

Dedicated to my Mother.......
Late Mrs. Laxmi Semalty
(04 Nov. 1949- 21 Aug. 2019)

Preface

The pharmaceutical industry is one of the fast-growing industry globally. The industry requires the skilled persons to meet the demand of fast growth. To meet the demand of the industry, the Pharmacy Council of India revised the pharmacy syllabi and came up with a uniform national syllabi in 2014. In the PCI syllabi 5th semester B. Pharm a paper "Industrial Pharmacy-I" is there to specifically cater this need and to fill the gap between industry and academia. Though the paper may be shifted to any other semester in a particular institute or University, the syllabi is uniform nationally. Moreover, in M. Pharm industrial Pharmacy and Pharmaceutics many relevant topics of industrial Pharmacy are there. The present book is an exhaustive review of Semalty's Industrial Pharmacy-I book. This is an effort to present the Industrial Pharmacy-I subject in a very effective and unique way for quick revision.

After the huge success of our previous book "Essentials of Pharmaceutical Technology", we are happy to present my new work to the B. PHARM/M PHARM students, faculty members, researchers and industry personnel. The book is my second endeavour after presenting India's first Pharmacy MOOC at Ministry of India's SWAYAM platform. In the online course the entire syllabi have been covered in 12 weeks, in 40 lectures of 30 minutes each. The course is available for credit transfer through SWAYAM MOOC for students. We got a huge response across the nation for our online course. More than 7500 learners are enrolled in the course across many countries. The course runs in every July - Dec semester. I welcome all the learners to join the free online course at SWAYAM platform (www.swayam.gov.in).

The book Semalty's Industrial Pharmacy-I covered all the important aspects of industrial Pharmacy including preformulation, tablets, capsules, liquid orals, parenteral, aerosols, ophthalmic, cosmetics and packaging. The content has been evaluated by the experts assigned by SWAYAM at national level for the quality. The content is power packed with easy to learn presentation for readers. Semalty's Reviews on Industrial Pharmacy-I provide an excellent opportunity to revise the entire syllabi of Industrial Pharmacy-I in a very effective manner. Besides, the book would be very helpful to present a ready reference for Industrial Pharmacy subject paper and to prepare for GPAT or other pharmacy competitive exam. The book provides to the point presentation, MCQs, glossary, FAQs for each chapter of Semalty's Industrial Pharmacy-I with student friendliness. Quick notes are the USP of the book.

I hope the book will be useful for all the stakeholders of pharmacy academia and industry. I welcome suggestions for improvement and feedback about the book.

- ***Dr Ajay Semalty***

Acknowledgements

I would like to thank all the persons who helped in designing, planning and finishing this book. Firstly, I would like to thank my family members for their continued encouragement, support and assistance. Without them, it would not have been possible to complete the book.

As John Cotton Dana said "*Who dares to teach, must never cease to learn*", this is the teacher who is solely responsible for quality of teaching and learning process. In the process of teaching and learning, we also learned lot of many things about the likes and dislikes of the students regarding the choice of books and reading habits. I have tried to focus my efforts on those points, to make the book student friendly. I am thankful to our students whose input again helped in designing and planning this book.

The learning is a never-ending process that will positively enhance our awareness and knowledge and will positively contribute to our abilities as teachers. So, I am also thankful to all the respected teachers who read, refer and recommend this book to their students.

I warmly acknowledge the contribution of Dr. Mona Semalty, Assistant Professor, Department of Pharmaceutical Sciences, H.N.B. Garhwal University (A Central University) Srinagar (Garhwal) and Dr. Lokesh Adhikari, Senior Project Associate, UGC-DAE CSR project, Department of Pharmaceutical Sciences, H.N.B. Garhwal University (A Central University) Srinagar (Garhwal) for their vital contribution in many chapters of the book.

I humbly acknowledge Ministry of Education, Govt. of India, SWAYAM National Coordinator CEC, New Delhi, and the Team EMRC Roorkee for providing grant and vital support in development of Industrial Pharmacy online course.

I also thank Prof B. Suresh, President, Pharmacy Council of India, Prof. P.K. Sahoo, Director Research, DPSR University New Delhi, Prof. S.H. Ansari, Faculty of Pharmacy, Hamdard University, New Delhi; Prof. Pawan Kumar Dubey, Director, Vivekananda College of Pharmacy, Indore; Prof Munish Ahuja, Department of Pharmaceutical sciences, Guru Jambheshwar University of Science and Technology Hisar; Prof. Y. S. Tanwar, B. N. College of Pharmacy, B.N. University Udaipur (Raj.); Prof. Vandana Panda, K.M. Kundanani College of Pharmacy, Mumbai; Prof. N.S. Hari Narayana Moorthy, Dean Pharmacy, Indira Gandhi National

Tribal University, Amarkantak, and all the other well-wishers who always keep triggering me time to time for the work like this.

I am thankful to my project Associates (Mr Mukesh Pandey and Mr Himanshu Mishra) and all the staff members of Department of Pharmaceutical Sciences, H.N.B. Garhwal University Srinagar (Garhwal) for their moral support. I also extend my gratitude to the entire University family of Garhwal university for their continuous support.

I humbly acknowledge Mr. Anil Shah, and his entire Publishing team of PharmaMed Press Hyderabad, India, with special thanks to Mr. Naresh Davergave and his entire production team for their efforts, guidance and timely support in preparation of this work.

I am always indebted to my Guru, Prof M.S.M. Rawat, Former Vice Chancellor and currently working as Advisor, Department of Higher Education Government of Uttarakhand for his continuous support and blessings.

Last but not the least, I thank my mother Late Mrs. Laxmi Semalty and God almighty for giving strength, patience and determination to complete this work on time in present form.

- Dr Ajay Semalty

Contents

Chapter 1

Preformulation I
(Physical Form: Crystal & Amorphous)

Dr. Ajay Semalty [M Pharm, MBA, Ph D]
Department of Pharmaceutical Sciences
H.N.B. Garhwal University
Srinagar Garhwal-246174, INDIA

API/ Drug

"API or drug is a chemical entity with affinity to the receptors and with positive or negative or nil intrinsic activity."

- **Timing of Preformulation**

  After Preclinical trial

Before formulation development

- Or in case of developing a new dosage form of an existing API.
- API is never administered in a raw chemical form.
- Excipients + Drug (API) + Dosage form.
- Focus of Preformulation
- High degree of uniformity: in physical characteristics (weight. Content, hardness etc) and drug release.
- Physiological availability, Bioavailability
- Therapeutic quality.

Preformulation

"All the activities of characterization of physicochemical properties of the drug under study which are important to develop a stable effective and safe dosage form."

Challenges in Preformulation

- Very small amount of API available.

- In initial stage of drug discovery, we have only a very tiny amount of API available sometimes in few mg, that too impure.
- Only preliminary data like melting point, spectral data and structure is available.

Preformulation Properties

- ***Physical properties:*** Physical form (crystal & amorphous), polymorphism, particle size, shape, flow properties, solubility profile (pKa, pH, partition coefficient),
- ***Chemical Properties:*** Hydrolysis, oxidation, reduction, racemisation, polymerization
- BCS: dissolution & permeability

Planning Preformulation

- First identify the dosage form
- Pick and study the relevant physicochemical properties (as per the desired dosage form) and take it in priority.
- Consequences of poor PF
- A poor Preformulation study may lead to the disasters
- Unstable/ ineffective or less effective and unsafe dosage form
- Loss of development time
- Increased expenditure on development
- Triggering repeated need for in vivo bioavailability/bioequivalence studies
- Physical properties

Solid state is the most preferred state of API for developing any dosage form. **Why?**

- API can easily be crystallized
- Easily be purified (by crystalizing)
- Easy to handle than liquids
- Better chemical stability than that of liquids.
- Types of solids

Solids may be of three types as per the internal structure (Physical)

- Amorphous
- Liquid Crystal
- Crystal

Amorphous Form

- ***"These are the solids which do not exhibit long-range order in any of the three physical dimensions."***
- There may be short-range order for amorphous solids.
- The amorphous phase always show higher free energy, enthalpy, and entropy than the crystalline one.

Amorphous form	→	Small particle size	→
More hygroscopic	→	More surface area	→
More energy	→	Less stability	
(more reactivity)			

Use of Amorphous Form

- In improving solubility
- In improving the oral bioavailability of the poor water soluble drugs.

But these are **less stable** as compared to their crystal phase.

There exists **some amount of crystal form in amorphous forms** also **(Two state model : USP).**

- The challenge
- Stability issues
- They might change in to crystalline form with the passage of time
- Very hygroscopic
- Prone to hydrolytic degradation.
- Typical to be formulated
- Only few amorphous drugs containing dosage form are marketed and approved by FDA.

Some FDA approved amorphous drugs

- **Itraconazole,**
- **Nelfinavir mesylate**
- **Paroxetine**
- **Celecoxib**
- **Cefuroxime axetil**
- **Cefepodoximeproxetil**
- **Novobiocin**

So, the amorphous form should be avoided until the difference in solubility make a significant impact on bioavailability.

Liquid Crystals

"If the internal structure is having long-range order but only one or two dimensions"

- On the basis of number of components these can be further classified as single, binary, and ternary LCs.
- But these are not of much use in pharmaceuticals.

Crystal Form

- A majority of APIs are crystalline in nature.
- **"the solids with the internal structure having long-range order in all three dimensions."**
- The logical method of classification of crystal is based on **the angle between the faces.**
- If three dimensions are given by a, b, and c, then these crystals may be of several types on the basis of length of the faces and angle between these faces.
- If a=b=c and angle between all the faces is 90 degree it is called simple cubic crystal (three equal axes each at right angle)
- Different crystal forms

Crystal Habit

Moving to It is the relative development of different types of faces. Let's take the example of

- NaCl in aqueous solution → Crystallizes into → Cubic face
- NaCl in aqueous solution → Crystallizes in to → Octahedral faces (with small amount of Urea)

The crystal habit depend on the process, impurities or conditions. (It refers to the types of faces developed and not the shape of the faces)

Single Entity

"The ability of a substance to exist as two or more crystalline phases that have different arrangements and/or conformations of the molecules in a crystalline lattice is called polymorphism."

- If the drug forms a binary composite of crystalline lattice with another chemical it is called binary adduct. And so on like ternary …
- On the basis of the ionization states of these species the adducts may be ionic, molecular or ionic/molecular.

Cocrystals

- "Two or more molecules are hydrogen bonded to each other."
- The choice of **cocrystal formation, depends on the need**.
- **Caffeine-oxalic acid cocrystals** showed better stability even at high humidity (Tarsk et al. 2005).
- **Carbamazepine-sachharin cocrystals** showed better bioavailability, suspension stability and same stability as compared to its immediate release tablet (Hickey 2007).

Importance of Crystallinity

- Solubility of drug candidates can be altered by modifying the crystal form:
- Solubility can be improved by partial amorphization through developing adducts or binary composites of drug.
- **Onset of action** can be controlled by using the crystalline form.
- Crystalline form can delay the onset of action and prolong the drug release. If you mix both of the form you can modulate the release. **For example: lente insulin give quick action and prolonged release.**
- The purity standards are laid down by the properties of a pure crystal.

FDA States

"It is mandatory to establish whether or not the API being studied exist in more than one crystalline form. If yes, what are the properties of all different crystal forms. Like melting point, solubility, stability, safety and efficacy."

How Crystals Affect Solubility

- When a crystalline molecule is to be dissolved, firstly it is to come out from the crystal lattice. The amorphous solute molecule are free to move in a solvent so easily dissolved.
- THE ENTHALPY CONSIDERATION delays the entry of drug molecule from crystalline lattice to solvent.

Characterization

- **Melting point**
- **Capillary melting**
- **Hot stage microscopy**
- **Thermal analysis or Differential Scanning Calorimetry (DSC)**
- **X ray diffraction**
- **IR**
- **SEM**
- **Misc.: Synchrotron radiation, solid state raman spectroscopy, solid state NMR etc.**

Thermal Analysis

- **The most versatile and precise method.**
- **Most suited for preformulation (requires only 2-5 mg of sample).**
- In Differential Thermal analysis **(DTA)** sample is heated at a constant rate and difference of the temperature between the sample and a reference is measured as a function of temperature or time.
- In DSC, all things are same like DTA except additional measurement of enthalpy or energy required to keep the sample at same temperature as that of reference.

X Ray Powder Diffraction

X rays are EMR between UV and gamma rays. These are expressed in angstrom units.

When X rays are incident on crystalline solids, scattering of x rays takes place. This scattering is called diffraction.

This diffraction is unique for a single pure crystal.

And it is available in the repositories XRD data bank.

Bragg's law define the diffraction

$$n\lambda = 2d \sin \theta$$

Where λ is wavelength of a perfectly and monchromatic X-ray beam

θ is the angle of incident beam on crystalline sample

n is the order of reflection (an integer, usually 1)

d is the distance between planes in crystal

Concept of XRD

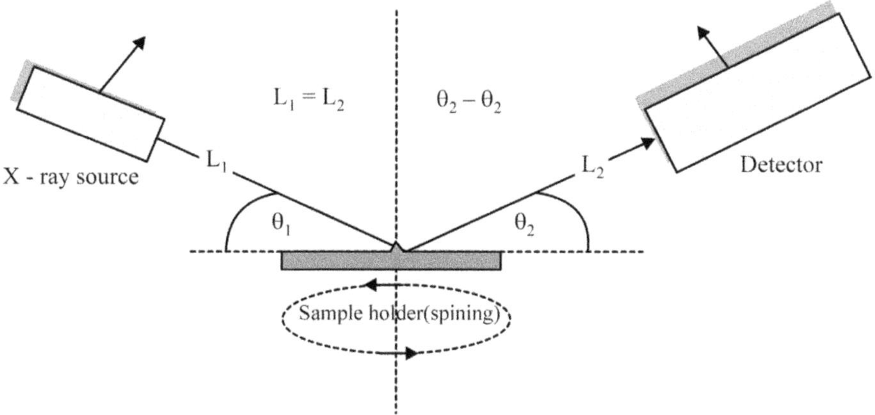

Infra Red Spectroscopy

- Provides qualitative information
- Different arrangement of atoms different molecular environment different stretching frequencies.
- Used to distinguish a polymorphic form.
- Functional groups, change, and interactions may be detected.
- Official books possess a data bank of IR of standard drugs.
- Aids to XRD in confirming the purity of molecule.
- Scanning Electron microscopy (SEM)
- The most advanced technique of visualizing the surface and also for particle size measurement.

Miscellaneous Method

- Synchrotron radiation,
- Solid State Raman Spectroscopy,
- Solid state NMR etc.

Take away Message

- Preformulation is the heart of formulation development
- Safety, stability and efficacy
- Amorphous form- unordered form, more soluble, high free energy, less stable
- Crystal form- defined shape, less soluble, more stable, less free energy
- Can be characterized by XRPD, IR, DSC and SEM

Chapter 2

Preformulation I

(Polymorphism, Particle Size & Shape)

Dr. Ajay Semalty [M Pharm, MBA, Ph D]
Department of Pharmaceutical Sciences
H.N.B. Garhwal University
Srinagar Garhwal-246174, INDIA

Lesson Plan

- Polymorphism
- Pseudopolymorphism
- Importance of polymorphism in preformulation
- Particle size and shape: Importance in preformulation
- Characterization of size and shape

Polymorphism

"The ability of a substance to exist as two or more crystalline phases that have different arrangements and/or conformations of the molecules in a crystalline lattice is called polymorphism."

Napoleon Army Story

(It was polymorphism not wrath of GOD...)

β-Tin or 'white' tin, stable above 18°C, Tetragonal, $I4_1$ /and a = b = 5.832, c = 3.182 Å; Metallic.

α-Tin or 'grey' tin, stable below 18° C, Cubic, Fd3m, a = b = c = 6.489Å, Non-metallic.

Types of Polymorphism

(a) On the basis of mechanisms crystal lattice formation

1. Packing polymorphism
2. Conformational polymorphism.

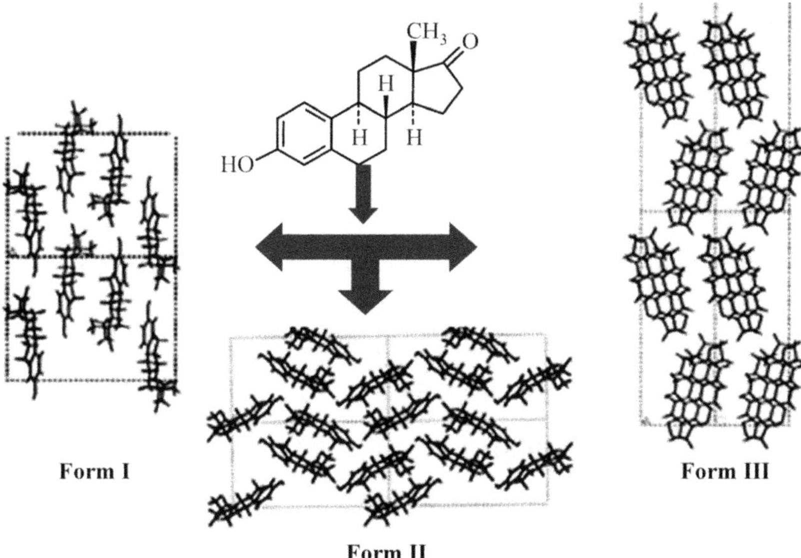

Figure 1. Estrone shows three polymorphic form (packing polymorphs).

(b) Thermodynamic classification

Depending upon whether or not one form can transform reversibly to another with respect to the change of temperatures and pressures, it can be classified into two forms

1. Monotropes
2. Enantiotropes

The temperature at which the two polymorphs have equal stability is defined as the transition temperature (*T*t).

If

Tt is located below the melting points of both polymorphs (Figure 2a) ---enantiotropic

Or

One polymorph can be reversibly changed into another one by varying the temperature or pressure. -- enantiotropic

Tt is located above the melting points of both polymorphs ((Figure 2b) --- monotropic,

Or

The change between the two forms is irreversible- monotropic

Pseudo Polymorphs

- "The phenomenon whereby solvent or water is incorporated in the crystal lattice or in interstitial voids, is termed as pseudo-polymorphism."
- Solvates (inclusion of the solvent of crystallization),
- Hydrates (inclusion of water of crystallization)

Residual solvents have been classified the ICH into three classes.

1. Class I solvents: Solvents to be avoided.
2. Class II solvents: Solvents to be limited.
3. Class III solvents: Solvents with low

Characterization Methods

As discussed in chapter 1, all the methods used for crystal characterization are used for polymorph form characterization. In characterization the complete profiling of all polymeric form are needed to be done.

Steps are as followed

1. Characterize the forms:

   e.g. - X-ray Powder Diffraction, - DSC / Thermal analysis, Microscopy, Spectroscopy
2. If the form shave different properties?(solubility, stability, melting point) then move to next step other no further testing required
3. If drug product safety, performance or efficacy is affected, then move to step 4, otherwise no further testing is required.
4. Set acceptance criterion for polymorph content in drug substance with respect to control on the ratio of forms, dissolution and stability.

Particle Size Reduction

- Mortar pestle:
  - for lab scale,
  - when material is less in quantity ;
  - more suitable or feasible in preformulation stage
- Ball milling:
  - for batch process,
  - in relatively large scale,
  - when material is in large quantity;
  - through a combined process of compact and attrition
  - consist of a hollow cylinder (containing different sizes of balls), rotated for grinding

  - critical factors include speed, mill size, wet or dry milling and amount of sample
  - but continuous milling may be detrimental for stability and crystallinity of the compound
- Micronization
  - Routine method for large scale size reduction by air jet milling
  - It involves feeding the sample into a confined circular chamber and then exposed to high velocity stream of air
  - Particles of less than 10 μm are obtained.
  - Solubility and bioavailability improves with the micronized drug e.g. felodipine
  - Changes in crystallinity may occur e.g. salbutamol sulphate

Particle Size Distribution

The uniform distribution of particles is also an important property. The particle size analysis can be done by following methods.

- Microscopy : Optical microscopy, Electron Microscopy
- Mechanical: Sieving method
- Electronic: Coulter counter
- Laser light scattering (DLS)
- **Miscellaneous techniques: Centrifugation, Air suspension, Sedimentation (Andreasen pipet, recording balance) etc.**

Chapter 3

Preformulation I

Solubility Profile (solubility, pH and pKa)

-Dr. Ajay Semalty
Department of Pharmaceutical Sciences,
H.N.B Garhwal University (A Central University)
Srinagar Garhwal-246174

Lesson Plan

- Concept of solubility
- Methods of improving solubility
- Characterization of solubility
- pH, pKa, concept and importance
- Determining/Characterization of pKa
- Importance, characterization and inter relation with respect to preformulation

"The extent of solute dissolved in a unit amount of solvent in certain conditions of temperature, and pH."

- Solubility

Solubility in Preformulation

Minimum requirement of API for preformulation is to determine Solubility

pKa

Minimum expected solubility?

Drugs with water solubility < 10 mg/ml (over the pH range of 1 to 7 at 37 °C) show the potential bioavailability problems.

The lower limit (below this absorption and bioavailability problem occurs) of Intrinsic dissolution rate is 1mg/cm2.min

Minimum expected solubility?

The limit of solubility is 1mg/ml.

If solubility < 1 mg/ml (Go for solubility improvement)

If salt form is not possible then liquid filling in soft or hard gelatin capsules remains the option.

Determining nature of drug and salt selection

Solubility in acid compared to aq. Solubility	Solubility in base compared to aq. Solubility	Nature of drug	Need of form
↑	-	Weak Base	pKacan be measured; salt should be formed
-	↑	Weak Acid	pKacan be measured; salt should be formed
↑	↑	Amphoteric or zwitterion	Two pKa,
-	-	Neutral Molecule	No measurable pKa; solubility manipulation by solvent or complexation

Using the Salt Form

The less soluble salts are intentionally used for specific reasons masking the taste (chloramphenicol palmitate); protecting the parent drug from excessive degradation in the gut, e.g. erythromycin Propionate; sustained release (propranolol laurate) etc.

Biopharmaceutical properties like dissolution rate, onset of action, duration of action, concentration at particular site, Cmax, Tmax and stability vary with the salt selection.

Modulating Solubility

- Use of co-solvent
- Hydrotrophy method
- By addition of polar group
- Use of solid solution
- Solid dispersion
- Eutectic mixture
- Micronization
- Use of surfactants
- Alternation of pH of solvent

- Use of suitable crystal form or metastable polymorphs
- Solvates formation
- Selective absorption on insoluble carrier
- Cyclodextrin complexation
- Phospholipid complexation

Intrinsic Solubility (S_0)

Solubility of a pure weak acid in acid and pure weak basic in alkali is calledIntrinsic solubility (So).

It is the fundamental solubility when completely unionized.

Calculating Intrinsic Solubility (So)

The solubility of the pure form is determined by phase solubility diagram.

The amount of drug to the amount of dissolving solvent is varied and solubility noted and recorded.

Impurity either increases or decreases the solubility. Impure form gives a non horizontal curve which intersects the y axis at a point which is actually the intrinsic solubility (Co).

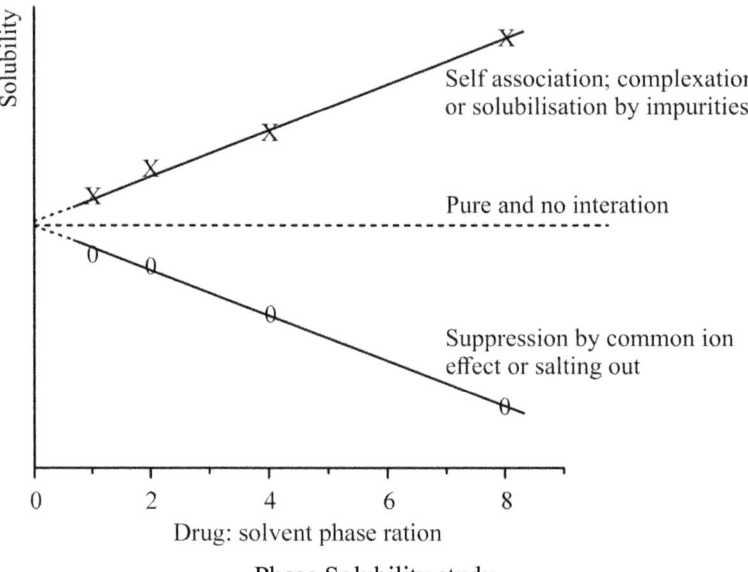

Phase Solubility study

Table 3 Solubility definition and classification (as per IP and USP)

Description	Parts of solvent required for one part of solute
Very soluble	<1
Freely soluble	1-10
Soluble	10-30
Sparingly soluble	30-100
Slightly soluble	100-1000
Very slightly soluble	1000-10,000
Insoluble	>10,000

Importance of pH in Preformulation

1. Injections should be in range of pH 3-9 to prevent tissue damage and pain at injection site.
2. Oral syrups can not be formulated too acidic for palatability reasons.
3. More alkali may attack the glass container.
4. If drug is susceptible to degradation in acidic pH, then its delayed release formulation is to be prepared.
5. The pH of formulation must not sensitize the site of application. e.g. pH for buccal application should in the range of 6.6 to 6.8.
6. GIT shows a variety of pH [pH 6.6 (buccal), pH 1.2 (stomach), pH 6.8 duodenum, pH 7-8 (small intestine)] throughout its length from oralcavity to colon. The dosage form should be stable at the pH of the intended or target site of absorption.

Ionization Constant (pKa)

Strong acids, e.g., HCl, are ionized at all pH values, whereas the ionization of weak acids is dependent on pH.

The extent of drug ionization at a certain pH depends on-

The relationship of pH and pKa

Smaller the acid/base dissociation constant weaker the acid or base.

$$K_a = [H^+]\ [A^-] / [HA] \qquad \text{..... for a weak acid}$$

Bigger the Ka stronger the acid

$$K_b = [Na^+]\ [OH^-] / [NaOH] \qquad \text{......for a weak base}$$

Henderson Hasselbalch Equation

Henderson Hasselbalch equation defines the ratio of the two species (unionzed and ionized)

Use of HH Equation

- To determine pKa with the change in solubility
- To predict solubility at a pH (If Co and pKa are known)
- To select suitable salt and its properties like pH and solubility.

"pKa value is the pH at which acidic or basic groups attached to molecules exist as 50% ionized and 50% nonionized in aqueous solution."

Determination of pKa

pKa value may be determined by

- Potentiometric titration,
- UV spectroscopy
- Solubility measurements
- HPLC techniques
- Capillary zone electrophoresis
- Foaming activity

Importance of pKa

The pKa value provides valuable data

- On the interaction of an ionizable drug
- On where the drug may be absorbed in the GIT
- To know how much to alter the pH to drive a compound to its fully ionized or nonionized form for analytical and other purposes, such as formulation, solubility, and stability.

Chapter 4

Preformulation I

Partition Coefficient & Flow Properties

-Dr. Ajay Semalty
Department of Pharmaceutical Sciences,
H.N.B Garhwal University (A Central University)
Srinagar Garhwal-246174

Lesson Plan

- Partition coefficient : Concept, Importance & Characterization
- Need to study flow properties
- Bulk density and (Carr's Index, Hausner's ratio)
- Method of determination of bulk density
- Angle of repose: Importance
- Determination of Angle of repose

Distribution or Partition Coefficient

The intrinsic properties of atoms or functional groups within the molecule give rise to the properties like lipophilicity, pKa, and aqueous solubility.

The lipophilicity of an organic compound is usually described in terms of a partition coefficient,

log P, which can be defined as the ratio of the concentration of the unionized compound,at equilibrium, between organic and aqueous phases:

Log P = [Unionized API] org / [unionized API] aq

Shake-Flask Method

The UVabsorbance of an aqueous solution is measuredbefore (ΣC) and after (C_w) being shaken with a known volume of octanol.

$$Ko/w = (\Sigma C - C_w)/ (C_w)$$

Partitioning Solvents

Many solvents are used for determining log P values.

On the basis of relative polarity to octanol the partitioning solvents may be of two types.

Hypodiscriminating solvents: Solvents more polar than octanol. E.g. butanol, pentanol; reflect more the lipophilicity of buccal region

Hyperdiscriminating solvents: Solvents less polar than octanol. E.g. oleyl alcohol, nitrobenzene, chloroform, Carbon tetra chloride, cyclohexane etc.; reflect more the lipophilicity of blood brain barrier

Preformulation and flow property

In preformulation stage the drug is available in small quantity so flow property is determined by bulk density and angle of repose.

Importance of flow Property in Pharmaceutical Industry

- For easy handling of drug powder in storage
- For easy handling in processing like flow through the hopper of tablet punching machine etc.
- For ensuring good mixing
- For ensuring content uniformity
- For ensuring the particle size uniformity and stability.
- Increase in crystal size or more uniform size leads to smaller angle of repose and smaller Carr's index

Bulk Density

First poured density and tapped densities are calculated and compared.

Carr's /Compressibility index (%) = [(Tapped density – poured density) / Tapped density] × 100

Table 1 Carr's index and type of flow

Carr's index (%)	Type of flow
5-15	Excellent
12-16	Good
18-21	**Fair**
23-35	**Poor**
33-38	Very Poor
>40	Extremely Poor

The fair and poor flow may be promoted using glidant like aerosil (0.2%).

Another scientist Hausner defined a new index:

Hausner Ratio = (tapped density/poured density) × 100

Hausner ratio < 1.25 → Good flow (= 20 % Carr)

Hausner ratio > 1.25 → Poor flow (= 33 % Carr)

USP provide the standard relation between these (**Table 2**).

Table 2 Relation between Carr's Index and Hausner Ratio (USP)

Carr's Index (%)	Flow property	Hausner ratio
≤ 10	Excellent	**1-1.11**
11-15	Good	**1.12-1.18**
16-20	**Fair- aid not needed**	**1.19-1.25**
21-25	**Passable- may hang up**	**1.26-1.34**
26-31	**Poor- must agitate / vibrate**	**1.35-145**
32-37	Very Poor	**1.46-1.59**
> 38	Very, very Poor	**>1.60**

Angle of Repose

"The angle of repose, or critical angle of repose, of a granular material is the steepest angle of descent or dip relative to the horizontal plane to which a material can be piled without slumping. At this angle, the material on the slope face is on the verge of sliding."

The angle of repose can range from 0° to 90°.

Table 3 Angle of repose and flow property (USP)

Angle of repose (Degrees)	Type of flow
25-30	Excellent
31-35	Good
36-40	**Fair- aid not needed**
41-45	**Passable- may hang up**
46-55	**Poor- must agitate / vibrate**
56-65	Very Poor
>66	Very, very Poor

If Angle of Repose in degrees is <20 then flow is excellent; 20-30 then Good; 30-34 Fair/passable; >40 then very poor

Table 4 Relationship between AR, Carr's index and type of flow

Flow	Angle of repose	Carr's index (%)
Excellent	< 25	5-15
Good	25-30	12-16
Fair to passable	30-40	18-21
Poor	> 40	23-35
Very Poor		33-38
Extremely Poor		>40

Chapter 5

Preformulation II

Hydrolysis, Oxidation, Reduction

-Dr. Ajay Semalty
Department of Pharmaceutical Sciences,
H.N.B Garhwal University (A Central University)
Srinagar Garhwal-246174

Lesson Plan

- Types of major chemical reactions
- Hydrolysis
- Oxidation
- Reduction
- Effect of temperature
- Effect of humidity
- Photolysis

Hydrolysis

It is the most common degradation reaction of any API. As water is a universal solvent, the hydrolysis is almost always associated with any formulation development directly or indirectly. Basically, for a drug in a solution, it is a nucleophilic attack of labile bonds by water.

"Hydrolysis is a two-stage process, where a nucleophile, such as water or the hydroxyl ion adds to, for example, an acyl carbon, to form an intermediate from which the leaving group breaks away in the second stage."

Factors affecting Hydrolysis

- Solution pH,
- buffer salts,
- ionic strength

- Use of nonaquoeus solvents
- presence of co-solvents,
- presence of complexing agent
- presence of surfactant
- temperature
- humidity

The Oxidation is

- **LEO: Loss of Electron is Oxidation**
- Removal of hydrogen atoms from a carbon atom or addition of an oxygen atom to a carbon atom.
- The most difficult reactions to understand,
- Difficult to prevent.
- Accelerated by the presence of trace metals
- Should be understood for the molecule under study and the formulation should contain suitable and compatible antioxidant or blend of antioxidants to prevent the same.
- Prevented by using antioxidants

Reduction

Reduction is the third most important chemical reaction after hydrolysis and oxidation.

Basically, reduction is indispensable with oxidation reaction. If one thing is oxidized the other reactant is reduced.

- **GER: Gain of Electron is Reduction**
- Removal of an oxygen atom from a carbon atom or addition of hydrogen atoms to a carbon atom is called reduction.
- Unlike oxidation which involves an increase in the oxidation number of an atom, the reduction involves decrease in oxidation number.
- An alkene upon reduction forms the corresponding alkane (by reacting with hydrogen).
- All the factors, which affect oxidation, also affect reduction.
- If oxidation is being prevented the same time, we are preventing the reduction.

Effect of Temperature

Arrhenius studied the effect of temperature on the reaction rate constant (k) and expressed their relationship by the following equation (Arrhenius equation).

$$K = A\ e^{-Ea/RT}$$

Where A = a constant which is termed as the frequency factor

or $$\log K = \log A - \left(\frac{Ea}{2.303RT}\right)$$

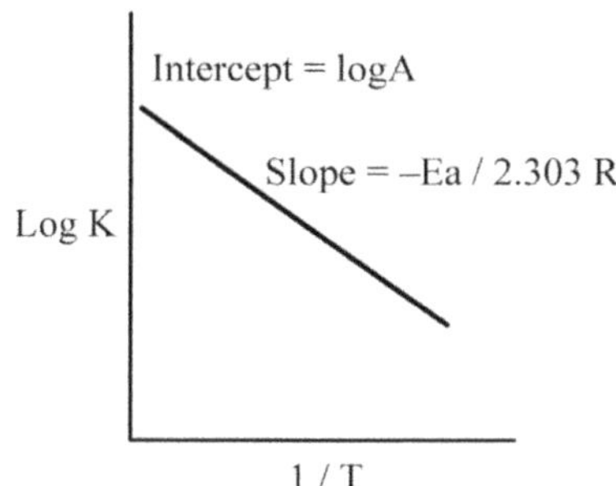

Arrhenius plot

Limitations of Arrhenius Equation

- Arrhenius equation was based on the assumption that the reaction mechanism does not change as a function of temperature, i.e., Ea is independent of T, but this is not true practically for every case. Therefore, it may not be valid for some reactions, especially the complex reactions.
- Higher temperature may evaporate solvents thus producing unequal moisture content at different temperatures.
- At elevated temperatures, there is less relative humidity and oxygen solubility, which makes it difficult to predict the stability of drugs (which are sensitive to the presence of moisture and oxygen) at room temperature.
- The viscosity of disperse systems is decreased at high temperatures, and thereby altering the physical characteristics.

Relative Humidity (RH)

Water vapor carrying capacity of air is called humidity."

When a piece of air of certain temperature and pressure is saturated with water vapors, then the humidity of that air is called saturation humidity or 100 % RH. In comparison to this the humidity is expressed in % RH.

Photolysis

Various drugs are sensitive to sunlight and upon exposure to the light these are degraded. For example, indomethacin, amphotericin B, riboflavin, tetracycline etc.

Avoiding Photolysis

Work in dark/ cover the apparatus by dark paper/film using

- amber glass bottles,
- cardboard outers
- aluminium foil overwraps
- blisters

Chapter 6

Preformulation II

Racemization

-Dr. Ajay Semalty
Department of Pharmaceutical Sciences,
H.N.B Garhwal University (A Central University)
Srinagar Garhwal-246174

Lesson Plan

- Stereochemistry and isomerism
- Chirality/Racemization: basic concept
- Examples
- Application

Stereochemistry and Isomerism

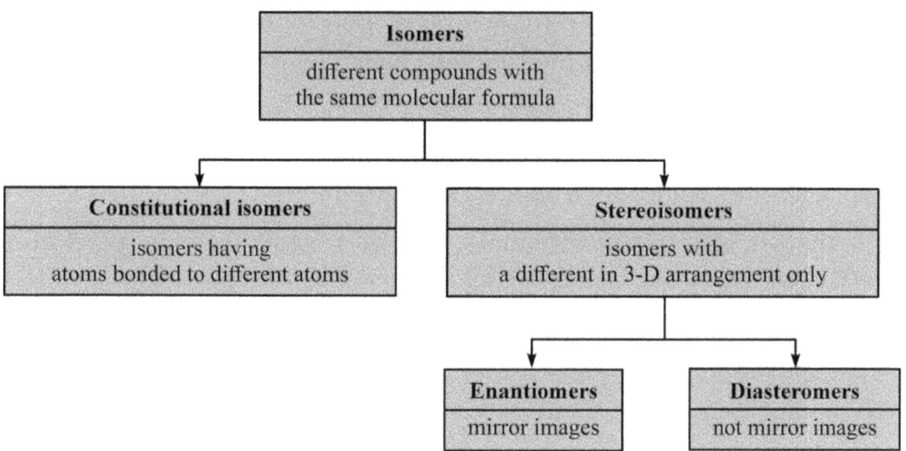

Mirror Image/ Superimposability

Two mirror images may or may not be superimposable. Like enantiomers first type of stereoisomerim that these **Enantiomers:** non superimposable mirror images **(all chiral)** while in case of **diastereomers** they are not mirror images (may or may not be chiral). This is the difference between the enantiomers and diastereomers

- "**Racemization** is a process that occurs when a compound undergoes a reaction in which the transformation produces an equal mixture of both possible enantiomers. When two compounds are classified as enantiomers of one another, it means they are **non-superimposable mirror images**."
- "the conversion of an **optically active** substance into an optically inactive mixture of equal amounts of the dextrorotatory and levorotatory."

R means laevo form and we can also denote with (-) and S means dextro (+) form.

S thalidomide is teratogenic. R form is useful and can be used...

- To treat some symptoms of Hansen's disease (Leprosy).
- To treat some types of cancer (trials going on)
- In the treatment of Erythema Nodosum Leprosum (ENL), a painful inflammatory dermatologic reaction of leprosy

Advantages of using Single Isomeric Products

- Improved therapeutic index will be there
- There will be Reduced drug interactions
- Less complex and more selective pharmacological profile will be there
- Less complex pharmacokinetic profile will be there

"the use of single enantiomers may be advantageous by permitting better patient control, simplifying dose-response relationships." - FDA

Chapter 7

Preformulation II

Dissolution, Permeability and BCS

-Dr. Ajay Semalty
Department of Pharmaceutical Sciences,
H.N.B Garhwal University (A Central University)
Srinagar Garhwal-246174

Lesson Plan

- Dissolution: Goals, basic concept and theories
- Dissolution apparatus
- Factors affecting dissolution
- Permeability; Concept
- Determining Permeability: Diffusion Cells
- BCS: Concept
- Class of drugs and examples
- Application of BCS

Dissolution

"Dissolution is defined as the process by which a known amount of drug substance goes into solution per unit of time under standardized conditions."

Dissolution testing is an official test recommended by all pharmacopoeias (viz. official books) for evaluating drug release of solid and semisolid dosage forms.

Dissolution: Concept

It is a two steps process

Step 1- Molecules are released from dosage form to media creating a saturated layer/stagnant layer, adjacent to the solid surface.

Step 2- Drug is diffused as per concentration gradient from higher concentration to lower concentration.

And it can be expressed with the help of Modified Noyes-Whitney's equation

$$\frac{dx}{dt} = \frac{A.D.K}{g} Cs - \frac{xd}{V}$$

- dx/dt is the dissolution rate,
- A is the surface area of the particle available for dissolution,
- D is the diffusion rate constant,
- K is the oil water partition coefficient,
- h is the thickness of the stagnant layer surrounding the particle,
- Cs is the saturation solubility of the drug,
- Xd is the amount dissolved of drug at time t and
- V is the volume of the dissolution media.

Dissolution: Theories

Dissolution can be explained with the help of some theories/ models. There are three major theories;

- Diffusion layer Model (Film Theory)
- Penetration or Surface Renewal Theory (Danckwerts' Model)
- Interfacial Barrier Model (Double Barrier Mechanism or Limited Solvation Theory)

Official Dissolution Test App. IP

These dissolution apparatus are standard, standard means each and every specification is final say for example in IP there are two types of official dissolution test apparatus (1) Paddle type apparatus is called type I apparatus and basket type apparatus is called type II apparatus.

Official Dissolution Test Apparatus USP

App. No.	Description	General Application
1	Rotating Basket	Oral IR and MR
2	Rotating Paddle	Oral IR and MR
3	Reciprocating Cylinder	Oral MR
4	Flow Through Cell	Oral MR
5	Paddle over disk	Transdermal DS
6	Cylinder	Transdermal DS
7	Reciprocating Holder	Transdermal DS or non-disintegrating oral MR

*IR = Immediate release, MR = Modified Release, DS= Delivery System

Dissolution Apparatus Factors

- Type of apparatus
- Dissolution Medium
- Agitation
- Temperature

Biopharmaceutical Classification System (BCS)

BCS is Biopharmaceutical Classification System (BCS). It is a drug development tool Proposed by Amidon et al. in 1995 based on aqueous solubility and intestinal permeability the drugs are classified into four major classes.

The rate and extent of drug absorption from solid oral dosage forms depend on three things

- dissolution,
- solubility and
- Intestinal permeability

Biopharmaceutical Classification System (BCS)

Depending upon solubility and permeability Amidon 4 box model, it is classified into four classes (**Table 5**);

Table 5 Biopharmaceutical Classification System (BCS)

Class	Solubility	Permeability
I	High	High
II	Low	High
III	High	Low
IV	Low	Low

BCS: Application

- For correlating the *in vitro* dissolution profiles with *in vivo* bioavailability of drugs
- For developing the *in vitro* dissolution specification
- For developing a strategy for improving the bioavailability of new chemical entities
- For predicting drug bioavailability: solubility or permeability limited.

Chapter 8

Preformulation II

Polymerization

-Dr. Ajay Semalty
Department of Pharmaceutical Sciences,
H.N.B Garhwal University (A Central University)
Srinagar Garhwal-246174

Lesson Plan

- Polymerization: basic concept
- examples
- application

Classification of Polymers

We can classify in many ways. A common classification is on the basis of origin of these polymers.

- Natural: Cellulose, guargum, xanthan gum, chitosan, alginate etc.
- Semisynthetic: Hydroxy ethyl cellulose, HPMC
- Synthetic: SCMC, CA, CAP, MC, EC, PVA, PVP etc.

Other basis of the classification

- ***Structure of the Monomer Chain***
- ***Polymerization***
- ***Monomers***
- ***Molecular Forces***
- ***Biodegradability***

Drug targeting is the ability of the drug to accumulate in the target organ or tissue selectively and quantitatively, independent of the site and methods of its administration.

USP defines *"Modified release dosage forms'* as"one for which the drug release characteristics of time course and/or location are chosen to accomplish therapeutic objectives not offered by the conventional dosage forms."

Chapter 9

Tablets-I

Introduction

-Dr. Ajay Semalty
Department of Pharmaceutical Sciences,
H.N.B Garhwal University (A Central University)
Srinagar Garhwal-246174

Lesson Plan

- Definition and ideal properties
- Advantages, Disadvantages, and Classification of tablets.
- Types and classification of excipients
- Basic steps of Formulation of tablets,
- Granulation methods: process, factors and equipment
  - Wet Granulation (WG)
  - Dry Granulation (DG)

Tablets

Tablets are the most common and convenient dosage form. About 70% of the total medicines are in Tablet form.

"Tablets are solid, flat or biconvex dishes, unit dosage form, prepared by compressing a drug or a mixture of drugs, with or without diluents."

Ideal Characteristics of Tablets

A tablet should have the following ideal properties.

- Elegant product identity
- Free of defects like chips, cracks, discoloration, and contamination
- Sufficient strength (to withstand mechanical shock and agitation during its production, packaging, shipping and dispensing.

- Chemical stability: shelf life
- Physical stability: color, physical integrity like shape etc.
- Release the drug in desired and reproducible manner.

Advantages of Tablet Dosage Forms

Tablet has a number of advantages, one of the major advantages over capsule, is that tablet is an essentially tamperproof doses form. Other potential advantages of tablets are as followed:

- These are the unit dosage forms which offer the greatest flexibility (with respect to doses) of all oral doses forms.
- Its cost is lowest of all oral dosage forms.
- These are the lightest and most compact of all oral doses forms.
- These are in general the easiest and cheapest to package and ship of all oral dosage forms.
- Product identification is simplest and cheapest, requiring no additional processing steps.
- These may provide the greatest ease of swallowing with least tendency for "hang-up' above stomach, especially when coated, provided that tablet disintegration is not excessively rapid.
- These are better suited to large scale production than other unit oral forms.
- These have best combined properties of chemical, mechanical, and microbiological stability of all the oral forms.
- Objectionable odor and bitter taste can be masked by coating technique.
- Generally, more stable than liquids with longer expiration dates.
- Release rate of the drug from tablet can be tailored to meet pharmacological requirements.

Disadvantages of Tablet Dosage Form

- Drugs which are destroyed in stomach cannot be prepared in tablets e.g proteins/peptides etc.
- Drugs with bitter taste or odour cannot be formulated in tablets without modification (capsules may be tried)
- Peak and valley effect in plasma time concentration curve for IR dosage forms
- The drug with too short half lifecan not be formulated.

Classification of Tablets (USP)

I. Immediate-release tablet

- Disintegrating tablet (conventional tablet)
- Chewable tablets
- Effervescent tablets
- Sublingual and Buccal tablets
- Lozenges

II. Modified-release tablet

- Extended-release
- Delayed-release

Excipients

"An excipient is an inactive substance used as a carrier for the active ingredients of a medication".

The excipients are selected on the basis nature of drug, nature of other excipients, compatibility with drug and other excipients and the desired release or performance characteristics of tablet.

Classification of Excipients

A. **Primary Excipients:** The primary excipients impart satisfactory compression characteristics to the formulation. E.g. diluents, binders, glidants and lubricants

B. **Secondary Excipients:** The secondary excipients give additional desirable physical characteristics to the finished tablet. e.g. disintegrants, coloring, flavoring and sweetening agents.

Manufacturing Tablet

Tablet is the most complex Dosage form. The tablet manufacturing is an art as well as a science. It consists of the two important steps:

Step1: Blending of drug with excipients (mixing and granulating) and;

Step 2: Compression (by punches onto die cavity)

Objectives of Granulation

Conversion Powder to granules is granulation and it is performed with the following objectives.

- To assure an even distribution of the active compound in the final tablet

- To ensure content uniformity due to narrow size range
- To increase the density for easy storage and shipment
- To improve flow and compaction

Mechanism of Granulation

The granulation occurs due to following events.

- Wetting
- Nucleation
- Coalescence/ growth
- Consolidation
- Attrition/ breakage
- By solid bridges/ chemical reaction/ sintering/ crystallization/ deposition of particles/ adhesive and cohesive forces of High viscous binders

Wet Granulation

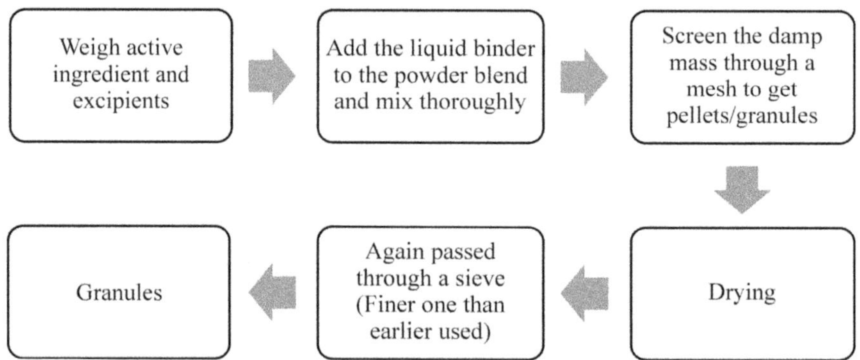

Figure 1 The wet Granulation Process

Fluidized Bed Granulation (FBG)

FBG is a multiple step wet granulation process. It is All in one process: pre-heat, granulate, and dry the powders. It allows close control of the granulation process.

Factors affecting Granulation

- Blend factors: Particle size of the drug and excipients
- Binder factor: Type, Volume and concentration of binder
- Granulation process factor: granulation time, equipment and drying rate

Dry Granulation

Dry granulation (DG) is done for moisture and heat sensitive products and the products which are not compressed well after WG. It is done without the use of liquid binder.

Process of DG

- Milling (Cutter Mill)
- Mixing (Sigma Blade Mixer)
- Slugging
- Screening (Oscillating Granulator)
- Mixing with Disintegrant and Lubricant
- Compression

Equipment for Granulation

- Roller compaction, extrusion spheronization, spray drying equipment
- Low shear/ High shear granulators
- Fluidized Bed Dryer (FBD)
- Continuous spouted bed
- High shear granulator (with the help of counter current operating blades) E.g. RMG (Rapid Mixer Granulator).

New Advanced Techniques

- Steam Granulation,
- Melt Granulation Granulation,
- Moisture Activated Dry Granulation (MADG),
- Moist Granulation Technique (MGT),
- Thermal Adhesion Granulation Process (TAGP),
- Extrusion-spheronization (ES),
- Continuous twin screw wet granulation (cTSWG)

Chapter 10

Tablets-II

Manufacturing Tablets

-Dr. Ajay Semalty
Department of Pharmaceutical Sciences,
H.N.B Garhwal University (A Central University)
Srinagar Garhwal-246174

Lesson Plan

- Basic preparation method of Tablet
- Tablet punching (equipment & process)
- Concept of compression
- Processing problems/ Tablet defects

Tablets Manufacturing

Tablet manufacturing is all about MGC.

- Mixing (drug and excipients)
- Granulation
- **Compression (Tablet punching)**

Tablet Punching Machine

A standard tablet compression machines consist of following components.

- **Hopper**: for temporary storing the material for compressing.
- **Feed frame**: for distributing the materials into the dies.
- **Dies**: for controlling the size and the shape of the tablet.
- **Punches**: for compressing the materials within the dies.

The Compaction Cycle

A typical complete tablet manufacturing cycle consists of the four steps:

(i) Die filling,

(ii) fill weight adjustment,

(iii) Tablet compaction and

(iv) Tablet ejection (from the die).

Types of Tablet Press

The tablet presses may be of various types.

- Single punch: One pair of punch and a die
- Rotary punching machine (16 station, 49 station)

Compression and Consolidation in Tablet Manufacturing

Tablet formation is one of the most complex process which involves the volume reduction of a blend of particles/granules (compression) followed by consolidation (into a defined solid dosage form/ tablet). So, tablet manufacturing or the tablet compaction consists of two steps

Step I: Compression (volume reduction)

Step II: Consolidation (shaping in tablet)

The process of compaction of a powder include following steps:

1. Particle rearrangement
2. Elastic, viscoelastic and plastic deformation of particles
3. Fragmentation of particles
4. Formation of interparticulate bonds

Tablet Defects

- **Capping, lamination, picking sticking, mottling, weight variation and double impression are the major tablet defects.**

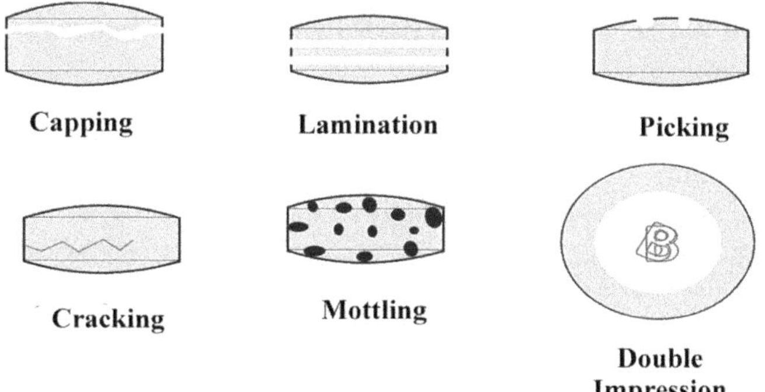

Chapter 11

Tablets-III

Tablet Coating

-Dr. Ajay Semalty
Department of Pharmaceutical Sciences,
H.N.B Garhwal University (A Central University)
Srinagar Garhwal-246174

Lesson Plan

- Tablet coating: Types
- Coating materials, formulation of coating composition,
- Methods of coating,
- Equipment employed and
- Defects in coating: Definitions, causes and remedies

Tablet Coating

- **P**rotection of the drug from the surrounding (environment) (air, light and moisture) and thus improve stability.
- **M**odifying drug release, as in enteric coating and extended-release formulation.
- **M**asking unpleasant taste or odour of the drug & batch differences in the appearance of raw materials
- **A**ppearance: Improving product **a**ppearance
- **I**dentification: Facilitating rapid **i**dentification by the manufacturer, the pharmacist and the patient (mostly colored).
- **S**trength: Increasing the mechanical **s**trength of the product.

Types of Tablet Coating Processes

Tablet coating may be of three major types.

- **Sugar coating**
- **Film coating**
- **Compression coating**

Steps for Sugar-Coating Process

- Sealing/waterproofing (of the tablet cores)
- Subcoating
- Smoothing
- Coloring
- Polishing
- Printing

Film Coating

"It involves the deposition, usually by spraying method, of a thin uniform film of a polymer formulation surrounding a tablet."

Types of Film Coating

1. *Immediate-release (non-functional) film coating*: They do not affect the biopharmaceutical properties of the tablet. They are readily soluble in water.
2. *Modified-release (functional) film coating*: They allow the drug to be delivered in a specific manner; i.e. they affect drug release behavior. Modified release film coatings are sub-classified into;
    - *Delayed-release coating (enteric coating)*:Only soluble in water at pH $\geq$ 5-6. Intended to protect the drug from gastric acidic pH (for acid labile drugs). Used for colonic drug delivery systems.
    - *Extended-release coating*: Mostly water-insoluble. Designed to ensure consistent drug release manner over a long period of time (6-12 hr) and thus decreasing dosing regimen and improving patient compliance.

Composition of Film Coating Agents

Film coating formulation/ materials are composed of

- Polymer,
- Plasticizer,
- Colorants,
- Solvent (vehicle)

Film Coating Equipment

The conventional "coating-pan" and "fluid-bed coating equipment" are both used for coating.

Coating-Pan

Tablet coating system combines several components:

- a coating pan,
- a spraying system,
- an air handling unit,
- a dust collector, and
- the controls.

Fluidized Bed Coating

- Fluidized Bed Dryers (FBDs) are used in granulation, drying and coating.
- Wurster (1959) first described the application of sugar or film coatings to tablets suspended in an air stream.
- Heater, perforated basket, air and coating spray gun, filter bags.

Compression-Coating of Tablets

Although less popular, it gained increased interest in the recent years for creating modified-released products.

It involves the compaction of granular materials around a preformed tablet core using specially designed tableting equipment.

Tablet Coating Defects

- Twinning
- Cracking
- Blistering
- Chipping
- Cratering
- Picking
- Pitting
- Blooming
- Blushing
- Color Variation
- Infilling
- Orange peel.

Chapter 12

Tablets-IV

QC of Tablets

-Dr. Ajay Semalty
Department of Pharmaceutical Sciences,
H.N.B Garhwal University (A Central University)
Srinagar Garhwal-246174

Lesson Plan

- QC of tablets:
  - Apparatus,
  - Official and unofficial Methods
- Graphical presentations of Evaluation study

QC/ Evaluation of Tablets

- Official test
  - Dissolution
  - Content uniformity (weight variation)
  - Disintegration
- Unofficial test
  - Mechanical strength
  - Tensile strength
  - Friability Test for Tablets (now considered official)
  - Tablet size and shape
  - Thickness
  - Average weight

Dissolution

"Dissolution is defined as the process by which a known amount of drug substance goes into solution per unit of time under standardized conditions."

The test is performed

- for characterizing the bio-pharmaceutical quality of a product at different stages in its lifecycle.
- For choosing between different alternative formulations for further development

"Dissolution testing is an official test recommended by all pharmacopoeias for evaluating drug release of solid and semisolid dosage forms."

Need of Dissolution Testing

- To help assure lot-to-lot uniformity
- USP requirement applies even if there is no correlation with *in vivo* data because it is a very discriminating and useful control over manufacturing variables.
- To guide formulation development
- To help establish stability/expiration dates
- To demonstrate to FDA that products after scaleup and post-approval changes are bioequivalent to the originally approved product.
- When correlated with *in vivo* data, dissolution may be used instead of human biostudies.
- Used without *in vivo* correlation for minor changes.

Dissolution Test (U.S.P.)

For dissolution test Pharmacopoeias specify dissolution test medium and volume,

- type of apparatus to be used,
- rpm of the shaft,
- time limit of the test and
- assay procedure

Three Stage Acceptance Criteria (USP)

The USP prescribes the following three stage criteria of acceptance of dissolution data (Table 6).

Table 6 Three stage acceptance criteria (USP)

Stage	Number Tested	Acceptance Criteria
S_1	6	Each unit is not less than Q + 5%.
S_2	6	Average of 12 units (S_1 + S_2) is equal to or greater than Q and no unit is less than Q – 15%.
S_3	12	Average of 24 units (S1 + S2 + S3) is equal to or greater than Q, not more than 2 units are less than Q – 15%, and no unit is less than Q – 25%.

***Q*, is the amount of dissolved active ingredient**

Weight Variation (USP/IP)

This test is used for assessing batch to batch uniformity. It is applicable when the tablet contains 50 mg or more of the drug.

Table 7 Acceptance Criteria of weight variation (IP/BP/USP)

Average weight of tablet (According to IP/BP)	Limit	Average weight of tablet (According to USP)
80 mg or less	± 10%	130 mg or less
More than 80 mg or less than 250 mg	± 7.5%	130 mg to 324 mg
250 mg or more	± 5%	More than 324 mg

Content Uniformity

The test is performed for all coated and uncoated tablets and all capsules intended for oral administration with the size range of the dosage form 50 mg or smaller sizes.

Acceptance Criteria (IP/USP)

30 tablets are selected and 10 are assayed individually. At least 9 must assay within ±15% of the declared potency and none may exceed ± 25%.

If two or three of the individual values are outside the limits ±15% of the average value and none is outside the limits ± 25%, repeat the test using another 20 tablets. The tablets comply with the test if in the total sample of 30 tablets not more than three of the individual values are outside the limits 85 to 115% and none is outside the limits 75 to 125% of the average value.

The disintegration test

"The disintegration test is a measure of the time required under a given set of conditions for a group of tablets to disintegrate into particles which will pass through a 10 mesh screen."

Table 8 Disintegration Test Acceptance criteria (IP) / as per Monograph

Type of capsule		**Disintegration time**
Uncoated Tablets		15 minutes
Coated Tablets	Film-coated	30 minutes
	other coated tablets	60 minutes
Enteric-coated Tablets	*0.1 M hydrochloric acid*	Should not disintegrate in 120 minutes
	mixed phosphate buffer pH 6.8	60 minutes
Dispersible and soluble tablets		Within 3 minutes
Effervescent tablets		5 minutes

Tensile strength/ hardness

It is a non-compendial method. "This is the force required to break a tablet in a diametric compression test." (Crushing strength).

Friability Test

The test has been included in IP 2014. So, now it has become official test.

"The test is designed to evaluate the ability of the tablet to withstand abrasion in packaging, handling and shipping."

Tablets Size and Shape

The thickness of individual tablets may be measured with a micrometer

Tablet thickness should be controlled within a ±5% variation of a standard value.

Chapter 13

Liquid Orals
(Syrups and Elixirs)

-Dr. Mona Semalty, Dr. Ajay Semalty
Department of Pharmaceutical Sciences,
H.N.B Garhwal University (A Central University)
Srinagar Garhwal-246174

Lesson Plan

- Introduction to liquid orals/ Classification of liquid orals /definitions
- Solutions /merits demerits/ Different types /
- Solubility (methods to enhance)
- Methods of preparation of solution/stability of solutions
- Syrups, components of syrups, method of preparations of syrups
- Elixir, components of elixir, method of preparation of elixirs
- Packaging and storage

Liquid orals are the pharmaceutical dosage form in liquid form to be administered orally either in the form of solution, suspension, emulsions, elixir and many more.

Classification of Liquid Orals

- Monophasic liquids
  - Solution Elixir
  - Syrup
  - Liquid drops etc.
- Biphasic liquids
  - Suspensions
  - Emulsions

Syrup

"In medical terminology, medicinal syrups are nearly saturated solutions of 85% of sugar in water in which medicinal substances or drugs are dissolved".

Elixirs

"Elixir are clear, flavored Oral Liquids containing one or more active ingredients dissolved in a vehicle that usually contains a high proportion of sucrose or a suitable polyhydric alcohol or alcohols and may also contain Ethanol (95 per cent) or a dilute Ethanol".

Suspensions

"Suspensions are Liquids containing one or more active ingredients suspended in a suitable vehicle. Suspended solids may slowly separate on keeping but are easily redispersed".

Emulsions

"Emulsions are Liquids containing one or more active ingredients and are stabilized oil-in-water or water-in-oil dispersions, either or both phases of which may contain dissolved solids".

Classification of solutions based on for the purpose it is to be used

- Oral
  - Syrups
  - Elixirs
  - Spirits
  - Linctuses
  - Drops
- In mouth & throat
  - Mouth washes
  - Gargles
  - Throat paint
  - Throat sprays
- On Body Surfaces
  - Collodions
  - Lotions
  - Liniments

- In Body Cavities
  - Douches
  - Enemas
  - Ear drops,
  - Nasal sprays

Selection Criteria for Solvents

Solvents are selected on the basis of following properties.

- Solubility
- Clarity
- Low toxicity
- Viscosity
- Compatibility with other ingredients
- Chemical inertness
- Palatability
- Odor, color
- Economy

Concentration of Syrup-

According to B.P: 67.7% W/W

According to USP: 85% W/V

Classification of Syrups

Medicated –Contains therapeutic agents in it

Non medicated "Syrups containing flavoring agents, but not medicinal substances are called non-medicated or flavored syrups"

Methods of Preparation of Syrups

- Solution with heat
- Agitation without heat
- Addition of sucrose to liquid medicament
- Percolation method

Elixirs

Elixir is clear, sweetened hydro-alcoholic solution. Alcoholic content vary from 10% to 12% and up to 40% Intended for oral use usually flavored to enhance palatability.

Usually less sweet than syrups and less viscous.

Advantages of Elixirs

- Better able to maintain both water-soluble and alcohol-soluble components in solution.
- Has a stable characteristic.
- Easily prepared by simple solution.

Disadvantages

- Less effective than syrups in masking taste of medicated substances.
- Contains alcohol, accentuates saline taste of bromides

Chapter 14

Emulsion-I

(Introduction Theories and Identification Tests)

-Dr. Mona Semalty, Dr. Ajay Semalty
Department of Pharmaceutical Sciences,
H.N.B Garhwal University (A Central University)
Srinagar Garhwal-246174

Lesson Plan

- Introduction, definitions as per IP, USP
- Classification
- Advantages & Disadvantages
- Uses (oral/systemic/local and other routes)
- Test for identification
- Different theories of emulsification

Emulsion can be defined as **a thermodynamically unstable system consisting of at least two immiscible liquid phases, one of which is dispersed as globule (dispersed phase) in the other liquid phase (continuous phase), stabilized by the presence of an emulsifying agent**

Oral Emulsions are **Oral Liquids containing one or more active ingredients and are stabilized oil-in-water dispersions, either or both phases of which may contain dissolved solids.** Solids may also be suspended in Oral Emulsions. Emulsions may exhibit phase separation but are easily reformed on shaking.

-Indian Pharmacopoeia

Emulsions are two-phase systems in which one liquid is dispersed throughout another liquid in the form of small droplets. If oil is the dispersed phase and an aqueous solution is the continuous phase, the system is designated as oil-in-water emulsion. If water or an aqueous solution is the dispersed phase and oil or oleaginous material is the continuous phase, the system is designated as a water-in-oil emulsion.

-USP

Theories of Emulsification

Droplets can be stabilized by three methods

- By reducing interfacial tension
- By preventing the coalescence of droplets.
    - By formation of rigid interfacial film
    - By forming electrical double layer.
- Film theory or adsorption theory.
- Viscosity theory.
- Wedge theory

Types of Emulsions

The emulsions can be of various types: Two phase, three phase, micro emulsions. Among the two phase systems O/W, W/O are the major types, while the three phase systems may be O/W/O and W/O/W. Micro and nano emulsions are other major types.

Uses of Pharmaceutical Emulsions

The Pharmaceutical Emulsions may be used for systemic as well as topical uses.

A. Systemic

Oral

- For Nutrition- triglycerides (e.g. vegetable oils)
- For masking unpleasant flavour

Parenteral

- For nutrition - triglycerides
- For delivery of drugs poorly soluble in water
- For site-directed delivery of drugs
- For the small droplet size,
- For comparable to chylomicrons
- For delayed-release drug release (e.g. i.m.)

B. Topical

- These may be o/w or w/o. These can be applied or rubbed (Viscous in nature).
- For emollient effect, greasiness (enhancing acceptability).

Radiopaque Emulsions

Emulsions as drug delivery systems

Identification Tests

- Dilution test
- Dye Test
- Cobalt Chloride Test
- Fluorescence Test
- Conductivity Test

Table 1 Methods of Determination of Emulsion Type

Test	Observation	Comments
Dilution	Emulsion can be diluted only with external phase	Useful for liquid emulsions only.
Dye test	Water soluble solid dye tints only o/w emulsions and reverse. (Microscopic observation)	May fail if ionic emulsifiers are present.
$CoCl_2$/ filter paper	Filter paper impregnated with $CoCI_2$ and dried (blue) changes to pink when o/w emulsion is added.	May fail if emulsion is unstable or breaks in presence of electrolyte.
Fluorescence	Since oils fluoresce under UV light, o/w emulsions exhibit dot pattern, w/o emulsions fluoresce throughout.	Not always applicable
Conductivity	Electric current is conducted by o/w emulsions, owing to presence of ionic species in water	Fails in nonionic o/w emulsions.

Chapter 15

Emulsion-II

(Formulations of Emulsions)

-Dr. Mona Semalty, Dr. Ajay Semalty
Department of Pharmaceutical Sciences,
H.N.B Garhwal University (A Central University)
Srinagar Garhwal-246174

Lesson Plan

- Formulation of emulsions
- Stability of emulsions
  - Sedimentation and creaming.
  - Thermodynamic instability (coalescence or cracking).
  - Phase inversion
- Additives for formulation of emulsions
- Methods of preparation and Evaluation of emulsions
- Packaging and labeling of emulsions

Formulation of Emulsions

Oil Phase & Aqueous Phase

Oil Phase

Selection of oil phase is important as it requires analysis of toxicity of the oil its consistency and possible chemical incompatibilities. Example includes Mineral oils and Edible vegetable oils

Aqueous Phase

Aqueous phase is water and the internal phase concentration of water will be determined by the dosage requirement and Consistency desired.

The other additives used are-

- Emulsifying agents
- Auxiliary emulsifiers.
- Antimicrobial preservatives
- Antioxidants

The Stability Problems of Emulsions

- Flocculation
- Cracking or coalescence
- Creaming and sedimentation
- Phase inversion
- Ostwald ripening

Additives for Formulation of Emulsions

- Emulsifying agents
- Auxiliary emulsifiers.
- Antimicrobial preservatives
- Antioxidants

Emulsifying Agents

Hydrocolloids

- Natural-Plant origin and Animal origin
- Semisynthetic

Surface active agents (SAA) (surfactants).

Finely Divided Solids

The Hydrophilic Lipophilic balance System (HLB)

An HLB number (1-20) represents the relative proportions of the lipophilic and hydrophilic parts of the molecule.

Theory of Emulsification

Emulsions are formed when the two phases are stabilized with the help of emulgents. The droplets can be stabilized by the following methods

A. By reducing interfacial tension
B. By preventing the coalescence of droplets.
    - By formation of rigid interfacial film
    - By forming electrical double layer.

Techniques of Emulsification

Techniques of Emulsification can be classified in following way.

A. Laboratory scale preparation techniques

- Continental or dry gum method
- Wet gum method
- Bottle or Forbes bottle method
- Auxiliary method
- In situ soap method

B. Large scale preparation techniques

- Emulsification by vaporization
- Emulsification by phase inversion
- Low energy emulsification

Equipment for Emulsification

- Mechanical stirrers
- Ultrasonifier
- Homogenizers
- Colloid mills

Quality Control Tests for Emulsions

Physical Parameters of Evaluation

- Phase separation
- Viscosity
- Globule size
- Electrical Conductivity
- Electrophoretic Properties

Stability Testing

- Phase separation
- Viscosity
- Electrophoretic properties
- Particle size and particle count

Microemulsions

Microemulsions can be defined as isotropic, thermodynamically stable solutions in which substantial amounts of two immiscible liquids (i.e., water and oil) are brought into a single phase by means of an appropriate surfactant or surfactant mixture.

Chapter 16
Suspension Dosage Form

-Dr. Mona Semalty, Dr. Ajay Semalty
Department of Pharmaceutical Sciences,
H.N.B Garhwal University (A Central University)
Srinagar Garhwal-246174

Lesson Plan

- Introduction of suspension, merits and demerits
- Classification
- Ideal properties of suspensions
- Theories of suspension
- Flocculated and deflocculated suspensions
- Formulation aspect of suspensions
- Evaluation, stability/packaging storage

Suspension

Suspension is heterogeneous system consisting of two phases. This biphasic dosage form is a coarse dispersion in which insoluble solid particles are dispersed in a liquid medium.

"Suspension is a dispersion of finely divided insoluble solid particles (disperse phase) in a fluid (dispersion medium). Dispersion medium is mostly aqueous or may be oily liquid, while dispersed phase comprises insoluble solids."

Advantages of Suspension

- Suspension can improve chemical stability of certain drugs.
  e.g. Procaine penicillin G
- Drug in suspension exhibits higher rate of bioavailability than other dosage forms.

- Duration and onset of action can be controlled.
  E.g. Protamine Zinc-Insulin suspension
- Suspension can mask the unpleasant/ bitter taste of drug.
  e.g. Chloramphenicol.

Disadvantages

- Physical stability, sedimentation and compaction can cause problems.
- It is bulky, sufficient care must be taken during handling and transport.
- It is difficult to formulate.
- Uniform and accurate dose cannot be achieved unless suspension are packed in unit dosage form.

Theories of Suspensions

Sedimentation Concept

These particles sediment according to stokes' law.

According to Stokes' law, the sedimentation velocity is expressed by following equation

$$v = \frac{2r^2(\sigma - \rho)g}{9\eta}$$

where v is velocity of sedimentation, a is the radius of particles, ρ_s density of solid particles, g is acceleration due to gravity and ρ is the density of liquid and η is the viscosity of the dispersion phase.

Interfacial Phenomenon

Zeta Potential

It indicates the degree of electrostatic repulsion. The high zeta potential (negative or positive) electrically stabilizes suspension while low zeta potentials tend to coagulate or flocculate the suspension.

DLVO Theory

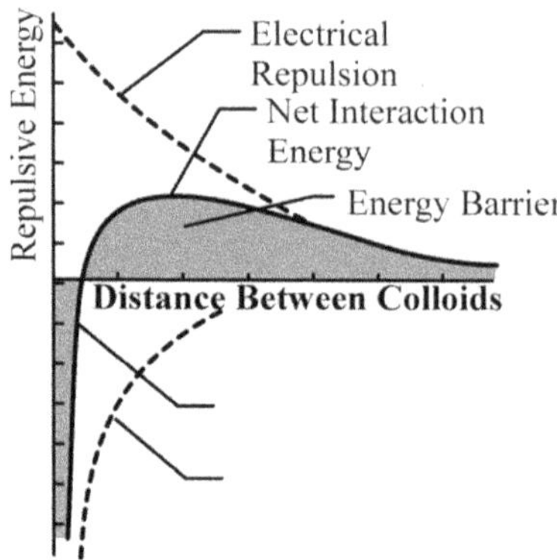

Figure 1 DLVO Theory: The net interaction curve.

Brownian Movement

Flocculated and Deflocculated Suspension

A suspension may be flocculated or deflocculated type depending on the relative magnitudes of force of repulsion, force of attraction between particles.

Flocculated System

In this type of system, the solid particles of dispersed phase aggregate leading to network like structure of solid particles in dispersion medium.

Deflocculated System

In this type of system the solid particles exist as separate entities in dispersion medium. The sediments form hard cake.

Formulation of Pharmaceutical Suspensions

Drug

Vehicle for Suspension

- Structured vehicles
- Other components

Viscosity Enhancers/ Suspending Agents

They are added with the objective to increase apparent viscosity of the continuous phase thus preventing rapid sedimentation of the dispersed particles. The selection of the type and concentration of a suspending agent depends on sedimentation rate of dispersed particles, pourability and spreadibility.

Method of Preparation

A. Precipitation Methods

(a) Organic solvent precipitation,

(b) Precipitation affected by changing the pH of the medium and

(c) Double decomposition.

B. Dispersion Methods

Evaluation of Suspension

Methods used for evaluation of suspension are categorized as :

(a) Sedimentation method

(b) Rheological method

(c) Electrokinetic method

(d) Micromeritic method

(e) Stability study

Nanosuspensions

Nanosuspensions are the recently developed biphasic dosage forms which have been developed for the delivery of poorly water-soluble and poorly water- and lipid-soluble drugs (belonging to class III as classified in Biopharmaceutical Classification System or BCS).

Nanosuspensions can be defined as colloidal dispersions of nano-sized drug particles that are produced by a suitable method and stabilized by a suitable stabilizer.

Chapter 17

Parenterals-I

(Introduction: Preformulation of Parenterals)

-Dr. Mona Semalty, Dr. Ajay Semalty
Department of Pharmaceutical Sciences,
H.N.B Garhwal University (A Central University)
Srinagar Garhwal-246174

Lesson Plans

- Introduction of parenteral and their principle characteristics
- Advantages and disadvantages of parenterals
- Routes of administration of parenterals
- Preformulation of parenterals: crystals characteristics, Chemical modification, polymorphism, pH &pKa

"The term parenterals is derived from Greek words "para" and "enteron", means outside the intestine." Parenteral products are prepared keeping in view to meet the Pharmacopoeialstandards for sterility, pyrogenicity, particulate matter etc.

"Parenteral articles are preparations intended for injection through the skin or other external boundary tissue, rather than through the alimentary canal, so that the active substances they contain are administered using gravity or force, directly into a blood vessel, organ, tissue, or lesion."

- United States Pharmacopoeia

Advantages of Parenterals

The parenteral formulations have various **advantages.**

- Quick onset of action
- Maximum bioavailability
- Suitable for the drugs which are not administered by oral route
- Useful for unconscious/vomiting/ Non-cooperative patients.
- Duration of action can be prolonged by modifying formulation.

- Suitable for nutritive like glucose & electrolyte.
- Suitable for the drugs which are inactivated in GIT or HCl (GI fluid)

Disadvantages of Parenterals

- Only trained person is required
- If given by wrong route, difficult to control adverse effect
- Difficult to save patient if overdose
- Sensitivity or allergic reaction at the site of injection
- Requires strict control of sterility & non pyrogenicity than other formulation.
- Once injected cannot be controlled (retreat)
- Injections may cause pain at the site of administration

Routes of Administration

Parenterals can be administered by various routes which can be classified as primary and secondary parenteral routes. Subcutaneous (s.c.), Intravenous (i.v.) and Intramuscular (i. m.) routes constitutes primary, while Intra-arterial, Intrathecal, Intraarticular, Intrapleural, Intracardial and Intradermal constitutes other routes of parenteral administration.

Preformulation of Parenterals

The important properties with reference to preformulation of parenterals includes

- solubility,
- pKa,
- pH,
- solid state characteristics,
- chemical modification of drug and
- Polymorphism etc.

Chapter 18

Parenterals-I
(Formulation of Parenterals)

-Dr. Mona Semalty, Dr. Ajay Semalty
Department of Pharmaceutical Sciences,
H.N.B Garhwal University (A Central University)
Srinagar Garhwal-246174

Lesson Plan

- Formulation of parenterals: basic steps of parenteral formulation
- Production of water for injection: Multi-effect still, vapour compression and reverse osmosis
- Formulation components: Vehicles (aqueous and non-aqueous vehicles) other added excipients like preservatives, solubilizers, wetting agents, buffers, antioxidants, chelating agents, suspending agents and tonicity adjuster etc.
- Methods of adjusting tonicity.
- Class I (Freezing point depression and Sodium chloride equivalent method)
- Class II (White-Vincent method and Sprowls method)

Formulation of Parenterals

The components of parenteral formulations are as followed.

- Therapeutic agents
- Vehicles
  - Water
  - Water miscible vehicles
  - Non- aqueous vehicles
- Added substances (Additives)
  - Antimicrobials
  - Antioxidants

- Buffers
- Bulking agents
- Chelating agents
- Protectants
- Solubilizing agents
- Surfactants
- Tonicity adjusting agents

Methods of Adjusting Tonicity

A. Freezing Point Depression

- freezing point depressions of 1w/v% drug solutions ($DT_f^{1\%}$)
- choose appropriate solute for adjusting tonicity
    - using $DT_{f,ref}^{1\%}$ determine required amount (w_{ref}) to cover remaining DT_f

$$w_{ref} = \left[\frac{0.52 - C\Delta T_f^{1\%}}{\Delta T_{f,ref}^{1\%}}\right] V_{req}$$

V_{req} = volume of water required

C = drug concentration (w/v%)

B. Sodium Chloride Equivalent (E) Method

NaCl (w/v%) = 0.90 – E × [drug] (w/v%)

E = NaCl equivalent

C. White-Vincent Method (USP Method)

calculates volume (V) in ml of isotonic solution that can be prepared by mixing drug with water/isotonic buffered solution

V = w × E × 111.1

w = wt. of drug (g)

Table 9 Classes and Examples of Parenteral Additives

Additives	Examples	Usual concentration (%)
Antimicrobials (Protect a preparation against the growth of microorganisms)	Benzalkonium chloride	0.01
	Benzyl alcohol	1-2
	Chlorobutanol	0.25-0.5
	Metacresol	0.1-0.3
	Butyl p-hydroxybenzoate	0.015
	Methyl p-hydroxybenzoate	0.1-0.2
	Propyl p-hydroxybenzoate	0.2
	Phenol	0.25-0.5
	Thimerosal	0.01
Antioxidants, Chelating agents, Buffers (Enhance the chemical stability of a solution)	Ascorbic acid	0.01-0.05
	Cysteine	0.1-0.5
	Monothioglycerol	0.1-1.0
	Sodium bisulfite	0.1-1.0
	Sodium metabisulphite	0.1-1.0
	Tocopherols	0.05-0.5
	Acetates	1-2
	Citrates	1-5
	Phosphates	0.8-2.0
	Salts of ethylenediaminetetraacetic acid	0.01-0.05
Protectants Prevents loss of activity of drug caused by stress or adsorption	Sucrose	2-5
	Glucose	2-5
	Lactose	2-5
	Maltose	2-5
	Trehalose	2-5
	Human serum albumin	0.1-1.0
Solubilising agents Increase and maintain drug solubility	Ethyl alcohol	1-50
	Glycerine	1-50
	Polyeyhylene glycol	1-50
	Propylene glycol	1-50
	Lecithin	0.2.0

Table 9 *Contd...*

Additives	Examples	Usual concentration (%)
Surfactants (Enhance the physical stability of proteins)	Polyoxyethylene	0.1-0.5
	Sorbitanmonooleate	0.05-0.25
Tonicity-adjusting agent Provide patient comfort by reducing pain and tissue irritation	Dextrose Sodium chloride	4-5 0.5-0.9

Chapter 19

Parenterals-I

(Types of Parenteral Preparations)

-Dr. Mona Semalty, Dr. Ajay Semalty
Department of Pharmaceutical Sciences,
H.N.B Garhwal University (A Central University)
Srinagar Garhwal-246174

Lesson Plan

- Drug injection: Injectable solutions
- Injectable emulsion, suspension and Dry powders
- Characteristics and properties of different types of parenterals
- Methods of preparations of injectable solution, dry powders, suspensions and emulsions

Types of Parenteral Preparations

1. Small volume parenterals
2. Large volume parenterals
3. Modified release (sustained release) injectables (SVIs delivered by routes other than IV)

Types of Parenteral Preparation (USP)

Solutions of drugs suitable for parenteral administration are referred as injection. According to USP injections are separated into

- [Drug] Injection: Liquid preparations that are drug substances or solutions thereof, e. g. Insulin.
- [Drug] *for* Injection: Dry solids that, upon addition of suitable vehicles, yield solutions conforming in all respects to the requirements for injections, e.g. Cefamandole Sodium
- [Drug] injectable emulsion: Liquid preparations of drug substances dissolved or dispersed in a suitable emulsion medium, e.g. Propofol

- [Drug] injectable suspension: Liquid preparations of solids suspended in a suitable liquid medium, e.g. Methyl Prednisolone Acetate.
- [Drug] *for* Injectable suspension: Dry solids that, upon the addition of suitable vehicles, yield preparations conforming in all respects to the requirements for injectable suspensions, e.g. Sterile Ampicillin for Suspension

Small Volume Parenterals

- Ampoules (solutions/ liquid filled concentrate)
- Vials
- Plastic blow-fill-seal ampoules
- Prefilled syringe
- Needle free injection

Table 10 Solutions V/S suspension injectables

Solutions	Suspensions

• The simplest and thus preferred form
• Risk of low stability of the active compound
• Normally rapid uptake
• Important quality parameters
  • pH
  • Osmolality (ionic strength)
  • Sterility
  • Content and impurities
	• Particles suspended in a solution
	• Not thermodynamically stable
	• Used for substances of low solubility or for controlled released formulations
	• Critical parameter the same as for solutions plus particle size

Syringeability refers to the handling characteristics of a suspension while drawing it into and manipulating it in syringe. It includes characteristics such as case of withdrawal from the container into the syringe, clogging and foaming tendencies, and accuracy of dose measurement.

The term **injectability** refers to the properties of the suspension during injection; it includes factors as pressure or force required for injection, evenness of flow, aspiration qualities, and freedom from clogging.

Chapter 20

Parenterals-I

(Plant layout for Parenterals)

-Dr. Mona Semalty, Dr. Ajay Semalty
Department of Pharmaceutical Sciences,
H.N.B Garhwal University (A Central University)
Srinagar Garhwal-246174

Lesson Plan

- Layout of the parenteral/sterile manufacturing area: Clean-up area, Preparation area, Aseptic area, Quarantine area, Finishing and packaging area
- Production facilities for parenteral/sterile product manufacturing: Principle, Personnel, Premises, Equipment
- Environmental/Air control: HEPA filters /Laminar airflow/Validation of HEPA Filters.

Different Areas for Production of Sterile Products

- Clean- up area
- Preparation area
- Aseptic area
- Quarantine area
- Finishing and packaging area
- **For the manufacture of sterile medicinal products 4 grades can be distinguished.**
- **Grade A:** The local zone for high risk operations, e.g. filling zone, stopper bowls, open ampoules and vials, making aspects connections. Normally grade A follows all the specifications for environment control.
- **Grade B:** For aseptic preparation and filling, this is the background environment for grade A zone.

- **Grade C and D:** Clean areas for carrying out less critical stages in the manufacture of sterile products.

Environment Control/Air Control

As we know that preparation of sterile product requires strict control over protection against microbial contaminations. Surrounding environment and air can be greatest source of contamination. It can be mi- nimised by using special bacterial and particulate filters. Air for aseptic area is then passed through high efficiency particulate air (HEPA) filters. HEPA filters are highly efficient it can remove 99.7% of all particles 0.3 μm or larger. A positive pressure is maintained relative to corridors, to prevent outside air from entering aseptic areas.

The HEPA Filters Validated by DOP-Dioctylphthalate-Method

When manufactured, DOP liquid plasticiser, heated to the point of vaporization and reconstituted into 0.3-micron particles to form a monodisperse aerosol. These single size particles are diluted with air until a conc. of 100 μg per litre is reached and aerosol mixture allow to be passed through the filter. The amount of penetration is measured on the downstream side with a forward light scattering photometer, giving the familiar reading of 0.03% or better studies were made of the effect of different size leaks on typical types of work such as photoresist, microweld, sterile transfers, and the like where it was shown that a leak of 0.01% was the border at which reject rates began to increase and excessive contamination was detected for in-situ DOP test this also represents the linear mass photometer threshold accuracy.

Air Classification Technique of Environmental Control

Air Classification system monitors the numbers of particle accepted and permitted in the manufacturing area (Table 11). (It is defined as the number of particles per cubic foot of air that are larger than 0.5 μm in diameter. e.g. Class 100, Class 1000, Class 10000). In the same way the air borne particulate classification system may be Grade A, B, C and D.

Table 11 Clean Room Classification

FS209 Cleanroom Classification	ISO 14644-1 Cleanroom Classification	NMT 0.5um Particle/m^3	Viable Microbes (cfu/m^3)	Average Airflow Velocity (fpm)	Air Change/hr
100,000	8	3,520,000	100	5-10	5-48
10,000	7	35,200	10	10-15	60-90
1000	6	35,200	7	25-40	150-240
100	5	3,520	1	40-80	240-480

Table 2 Air Borne Particulate Classification

	Maximum permitted number of particles/m^3			
Grade	At rest	In n operation		
	0.5 mm	5 mm	0.5 mm	5 mm
A	3500	0	3500	0
B	3500	0	350,000	2000
C	350,000	2,000	3,500,000	20,000
D	3,500,000	20,000	Not defined	Not defined

Module 21

Parenterals II

(Pyrogens and Pyrogenicity)

-Dr. Mona Semalty
-Dr. Mona Semalty, Dr. Ajay Semalty
Department of Pharmaceutical Sciences,
H.N.B Garhwal University (A Central University)
Srinagar Garhwal-246174

Lesson Plan

- Introduction to pyrogens/importance
- Endotoxins (classification and structure)
- Bioactivity of endotoxins
- Sources of pyrogen contamination
- Elimination of pyrogens
- Quality control test for parenterals

Pyrogens

Pyrogens are the metabolic products of microorganisms. Parenteral product must be free from pyrogens. If it is injected it causes increase in body temperature (causes fever), reduction of the blood pressure, shock, tissue damage, inflammation, thrombosis, hemorrhages of the skin and mucous membranes, sepsis, death.

Pyrogen Detection

Both the United States Pharmacopeia and the European Pharmacopoeia specify the rabbit pyrogen test and the Limulus amebocyte lysate (LAL) test.

There are 6 methods of pyrogen detection.

Method A. Gel-clot method: limit test

Method B. Gel-clot method: semi-quantitative test

Method C. Turbidimetric kinetic method

Method D. Chromogenic kinetic method

Method E. Chromogenic end-point method

Method F. Turbidimetric end-point method

Quality Control of Parenteral Includes

- Test for pyrogens: Rabbit test and Bacterial endotoxins test (BET).
- Particulate evaluation
- Leakers Test
- Sterility test

Test for pyrogens (Rabbit test)

I. Preliminary test (Sham Test)

II. Main test:

- Group of 3 rabbits
- Preparation and injection of the product:
  - Warming the product
  - Dissolving or dilution
  - Duration of injection: not more than 4 min
  - The injected volume: not less than 0.5 ml per 1 kg and not more than 10 ml per kg of body mass
- Determination of the initial and maximum temperature
- All rabbits should have initial T: from 38.0 to 39.8°C
- The differences in initial T should not differ from one another by more than 1°C

Table 12 Interpretation of results (Sham test)

No. of Rabbits	Individual Temp. rise (°C)	Temp. rise in group (°C)	Test*
3 rabbits	<0.6	≤ 1.4	Passes
If above not passes take additional 5 rabbits (for 3+5 = 8 rabbits)	<0.6 (not more than 3 should show ≥ 0.6)	≤ 3.7	Passes

*If above test does not pass the sample is said to be pyrogenic perform the test again

Limulus Amebocyte Lysate (LAL) test or Bacterial Endotoxins Test (Official in I.P. and B.P.).

LAL Test performance (The key points)

- Avoid endotoxin contamination
- Before the test:
  - interfering factors should not be present
  - equipment should be depyrogenated
  - the sensitivity of the lysate should be known
- Test:
  - equal volume of LAL reagent and test solution (usually 0.1 ml of each) are mixed in a depyrogenated test-tube
  - incubation at 37°C, 1 hour
  - remove the tube, invert in one smooth motion (180°), read (observe) the result
  - pass-fail test
    - If a firm gel is formed- Pyrogen present.
    - If an intact gel is not formed – Pyrogen absent.

Particulate Evaluation

Visual method (against black& white backgrond)

In light obscuration method equipment used to count and measure the size of particles by means of a shadow cast by the particle as it passes through a high-intensity light beam. BP, JP and USP limits for particulate matters determined by LOPC and MPC test is mentioned in Table 13 & Table 14.

Table 13 BP, JP and USP limits for particulate matters determined by LOPC test

Nominal Volume	≥ 10 μm	≥ 25 μm
More than 100 mL	25 particles/mL	3 particles/mL
100 mL or less than 100 mL	6000 particles/container	600 particles/container

In microscopic method sample is filtered through membrane filter (under ultra clean condition) and counting the particles on the surface of filter. If limit is exceeded for stage 1 the sample is subjected to stage.

Table 14 BP, JP and USP limits for particulate matters determined by MPC test

Nominal Volume	≥ 10 μm	≥ 25 μm
More than 100 mL	12 particles/mL	2 particles/mL
100 mL or less than 100 mL	3000 particles/container	300 particles/container

Leakers Test

Ampoules sealed by fusion are subjected for leakers test. Negative pressure is built within an incompletely sealed ampoule (submerged entirely in deeply colored solution, i.e. methylene blue solution). All leakers are discarded

Chapter 22

Parenterals II
(Sterility Test and Sterilization)

-Dr. Mona Semalty
-Dr. Mona Semalty, Dr. Ajay Semalty
Department of Pharmaceutical Sciences,
H.N.B Garhwal University (A Central University)
Srinagar Garhwal-246174

Lesson Plan

- Sterility tests: basics, compendial requirements, tests methods and results interpretation
- Sterilization/sterility
- Methods of sterilization
- Direct transfer method
- Membrane tests method
- Containers for parenterals
- Filling and sealing of ampules and closures system

Sterility and Sterility Test

Sterility means the complete absence of all viable microorganisms. It is an absolute term; that is, a product is either sterile or not sterile. Building sterility into a product through meticulously validated cleaning, filtration, and sterilization procedures is preferable than testing for sterility of a product subjected to marginal or inadequate production processes. The sterility test should never be employed as an evaluation of the sterilization process.

Sterility Test Methods

The USP and EP sterility tests specify two basic methods for performing sterility tests,

- The direct transfer (DT) or direct inoculation method and
- The **Membrane Filtration** (MF) method,

with a statement that the latter, when feasible, is the method of choice. In fact, in some cases, membrane filtration may be the only possible choice. IP, USP and EP provide the detailed SOPs of these tests.

Methods of Sterilization

"Sterilization can be defined as any process that effectively kills or eliminates transmissible agents (such as fungi, bacteria, viruses and prions) from a surface, equipment, foods, medications, or biological culture medium.

D-value

D-value is indicative of the resistance of any organism to a sterilizing agent. For radiation and heat treatment, D-value is the time taken at a fixed temperature or the radiation dose required to achieve a 90% reduction in viable count.

Z-value

Z-value represents the increase in temperature needed to reduce the D-value of an organism by 90%.

Methods of Sterilization

1. Physical Method
    - Thermal (Heat) methods
    - Radiation method
    - Filtration method
2. Chemical Method
3. Gaseous method

Dry Heat Sterilization

- Incineration
- Red heat
- Flaming
- Hot air oven

Moist Heat Sterilization

Moist heat may be used in three forms to achieve microbial inactivation.

- Dry saturated steam – Autoclaving
- Boiling water/ steam at atmospheric pressure
- Hot water below boiling point

Application of Filtration for Sterilization of Gases

- HEPA (High efficiency particulate air) filters can remove up to 99.97% of particles >0.3 micrometer in diameter.
- Air is first passed through prefilters to remove larger particles and then passed through HEPA filters.
- The performance of HEPA filter is monitored by pressure differential and airflow rate measurements.
- There are two types of filters used in filtration sterilization
    - Depth filters: Consist of fibrous or granular materials so packed as to form twisted channels of minute dimensions. They are made of diatomaceous earth, unglazed porcelain filter, sintered glass or asbestos.
    - Membrane filters: These are porous membrane about 0.1 mm thick, made of cellulose acetate, cellulose nitrate, polycarbonate, and polyvinylidene fluoride, or some other synthetic material. The membranes are supported on a frame and held in special holders. Fluids are made to transverse membranes by positive or negative pressure or by centrifugation.

Chapter 23

Capsule I

-Dr. Lokesh Adhikari, Dr. Ajay Semalty
Department of Pharmaceutical Sciences,
H.N.B Garhwal University (A Central University)
Srinagar Garhwal-246174

Lesson Plan

- Definition and basic concept of capsule dosage forms
- Advantages and disadvantages of capsules
- Capsule shell production
- Capsule shell storage
- Hard gelatin capsules

Capsule

Capsules are solid dosage forms in which the drug substance is enclosed within either a hard or soft soluble shell, usually formed from gelatin. The term capsule is derived from the Latin word *capsula*, meaning a small container.

Capsule

"Capsules are solid preparations with hard or soft shells of various shapes and capacities, usually containing a single dose of the active substance(s). They are intended for oral administration."

-British Pharmacopoeia

Capsule Shell Production

- Dipping
- Spinning
- Drying
- Striping
- Cutting
- Joining

Capsule Material

Gelatin

It is the major component of the capsule. Gelatin is non-toxic; soluble in biological fluids at body temperature, and it is a good film-forming material. Solutions of high concentration, 40% w/v, are mobile at 50°C as compare to other polymers like agar.

There are two main types of gelatin:

Type A: produced by acid hydrolysis of animal skins.

Type B: produced by basic hydrolysis of bovine bones.

The two types can be differentiated by their isoelectric points (7.0 – 9.0 for type A and 4.8 – 5.0 for type B) and by their viscosity and film forming characteristics.

- **Colorants**

  Two types of colorants are used: water soluble dyes (e.g. erythrosine) and pigments (e.g. iron oxides, titanium dioxide)

- **Preservatives**

  When preservatives are employed, parabens are often selected

Defects of Hard Gelatin Capsule Shells

Cracked: Body or caps with many splits

Double cap: Capsule with an additional cap covering the body

Double-dipping: Extra thick cap due to double-dipping

Failure to separate: Cap and body may not separate properly

Hole: Irregular opening in cap or body

Joined in the lock: Capsule in the locked position

Long cap/body: Length cap or body 1 mm more than specification

Mashed: Mechanically damaged, squashed flat

Pinched: Cap or body damaged in collets, pinch > 3 mm

Short body: Length 0.4 mm less than specification

Trimming: Whole or piece of trimmed end cap or body inside the capsule

Uncut cap/body: Untrimmed cap or body

Un-joined: Single cap or body

Dye speck: a Colored spot of pigment, different from cap or body-color

Chapter 24

Capsule II

-Dr. Lokesh Adhikari, Dr. Ajay Semalty
Department of Pharmaceutical Sciences,
H.N.B Garhwal University (A Central University)
Srinagar Garhwal-246174

Lesson Plan

- Introduction to hard gelatin capsules
- Methods of filling hard gelatin capsules
- Advancements in filling of hard gelatin capsules
- Formulation aspects of hard gelatin capsule

Hard Gelatin Capsules

These are also known as two piece capsules or dry powder capsules. More than 85% of hard capsule shells are made up of gelatin. Gelatin based capsules are elegant in appearance and easy to swallow. The medicament is protected from external environment with a faster release in comparison with tablet.

Hard gelatin capsules can be filled either by manual method or mechanical method. Any method of filling comprises of five necessary steps

- Feeding and rectification
- Separation
- Filling
- Joining
- Ejection

Hand-Operated Filling Devices

Automated Devices

Filling

Types of methods used for filling capsule shells

- Auger Fill Method
- Mechanical Vibration Filling Method
- Dosator Method
- Tamping Pin/Dosing Disc Method
- Compression Filling Method

Important Capsule Blend Properties

- Density
- Flow
- Adhesion
- Water content
- Compatibility
- Compressibility

Chapter 25

Capsule III

-Dr. Lokesh Adhikari, Dr. Ajay Semalty
Department of Pharmaceutical Sciences,
H.N.B Garhwal University (A Central University)
Srinagar Garhwal-246174

Lesson Plan

- Introduction to soft gelatin capsules
- Methods of filling soft gelatin capsules
- Advancements in filling of soft gelatin capsules
- Formulation aspects of soft gelatin capsule

Soft Gelatin Capsules

Definition

Soft capsules are a single-unit solid dosage form, consisting of a liquid or semi-solid fill enveloped by a one-piece hermetically sealed elastic outer shell. Formed, filled and sealed in one continuous operation, preferably by the rotary die process.

Composition

On the basis of formulational category

- Shell material (Gelatin/PVA/carrageenan)
- Plasticizer (single or a combination of plasticizers)
- Water as solvent
- Preservatives
- Colouring agents
- Opacifying agents
- Flavourings and sweeteners

On the basis of solubility

- Hydrophilic solvents
- Lipophilic materials
- Hydrophilic surfactants
- Lipophilic surfactant
- Cosolvents

Equipment

Product Pump for feeding the shell material

- Separate or attached
  Injection time adjustment for time regulated feeding
- Electronic or manual
  Injection wedge for shape creation
- Manual or pneumetic
  Rollers for filling material
- Electronic or pneumatic pressure based
  Ribbon system for film formation
- Electronic or pneumatic
  Oil quantity adjustment for core material feeding
- Electronic or manual
  Collection assembly for separating softgels
- Chute or divider
  Speed
- Fixed or variable
  Conveyer
- Unidirectional or bidirectional

Finished Product Tests

- Capsule appearance for elegance and better patient compliance.
- Active ingredient assay & related substances assay for uniformity of dosing and dosage form.
- Fill weight
- Content uniformity
- Microbiological testing for safety purpose.

Chapter 26

Capsule IV

-Dr. Lokesh Adhikari, Dr. Ajay Semalty
Department of Pharmaceutical Sciences,
H.N.B Garhwal University (A Central University)
Srinagar Garhwal-246174

Lesson Plan

- General acceptance criteria for capsules
- Evaluation of raw material
- Evaluation of in process material
- Evaluation of finished product

Evaluation for Raw Material

In this category, shell forming agent is prime focus of evaluation. Though, APIs are also tested for general tests of purity as per the monographs.

API testing involves

- FTIR,
- Assay,
- Limit test
- Or any other, if stated in monograph.

In case of shell forming agents, evaluation parameter includes

- Bloom strength (in case of gelatin),
- Viscosity (in solution form),
- Antimicrobial test etc.

Official tests for shell forming agents include

- Identification,
- Microbial contamination,
- Loss on drying,

- Isoelectric point,
- Conductivity etc.

Evaluation for in process material:

Moisture content:

Gel seal thickness:

Content of active ingredients

Uniformity of weight:

Uniformity of dosage units:

Table 15 Weight variation limit for capsules

Average weight of capsule contents	Percent deviation limit
Less than 300 mg	10
300 mg or more	7.5

Disintegration testing (not applicable to Modified-release Capsules)

Dissolution testing (like Tablets)

Chapter- 27

Pellets & Pelletization

-Dr. Lokesh Adhikari, Dr. Ajay Semalty, Dr. Mona Semalty
Department of Pharmaceutical Sciences,
H.N.B Garhwal University (A Central University)
Srinagar Garhwal-246174

Lesson Plan

- Introduction to pellet
- Methods of pelletization
- Equipment

Pellet has been described as a variety of agglomerates which are produced from diverse raw materials by using different pieces of manufacturing equipment.

Pellets are small, free-flowing, spherical particulates manufactured by the agglomeration of fine powders or granules of drug substances and excipients.

Pelletization Techniques

- **Powder layering**
- **Wurster coating process**
- **Solution/suspension layering**
- **Extrusion–spheronization**
    - Screw-fed extruders
    - Gravity-fed extruders
    - Ram extruders
- **Spherical agglomeration**
    - Liquid-induced
    - Melt-induced agglomerations
- **Cryopelletization**
- **Melt spheronization**

Chapter 28

Ophthalmic Preparations

Introduction, Absorption through Eye, Formulation Considerations

-Dr. Lokesh Adhikari, Dr. Ajay Semalty
Department of Pharmaceutical Sciences,
H.N.B Garhwal University (A Central University)
Srinagar Garhwal-246174

Lesson Plan

- Introduction to ophthalmic dosage forms
- Ocular route and absorption
- General properties of ophthalmic products

Definition

Three ways can do the instillation of the specialised dosage forms into the eyes:

1. topical (onto the external surface of the eye),
2. intraocular (administered inside) and
3. periocular (adjacent to the eye).

Sometimes these formulations are also used in conjunction with an ophthalmic device. These are known as ophthalmic dosage forms.

Ophthalmic dosage forms can be of the semisolid or liquid type, including solutions, suspensions, and ointments. Gel, gel-forming solution, ocular insert, intravitreal injections and implants are few newly developed forms of ophthalmic drug delivery.

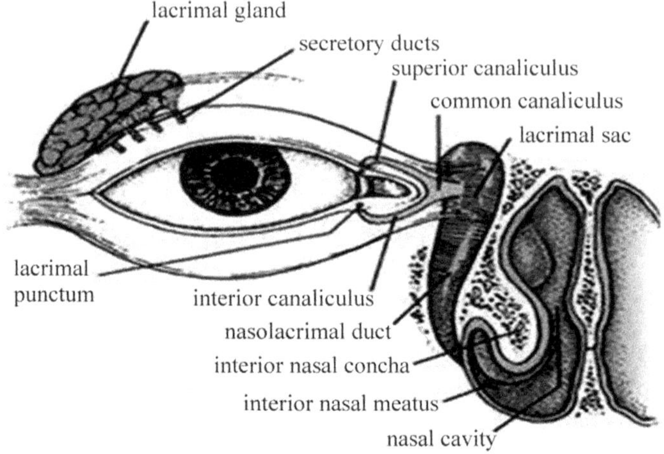

Figure 1. Anatomy of human eye.

Factors Affecting Drug Availability

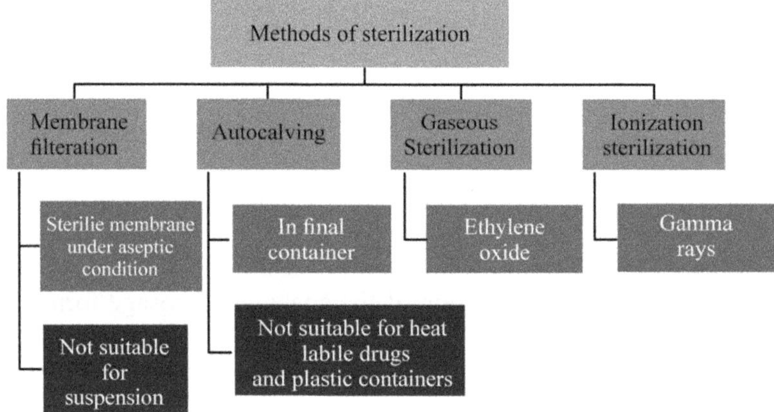

Table 16 Types of excipients of ophthalmic dosage forms

Category	Example	Properties
Cationic wetting agents	Benzalkonium chloride	• 0.01-0.1%
	Disodium edetate (EDTA)	• Ability to render the resistant strains of PS aeruginosa • More sensitive to benzalkonium chloride
Organic mercurial	Phenylmercuric nitrate	• 0.002-0.004%
	phenylmercuric acetate	• 0.005-0.02%.
Esters of p-hydroxybenzoic acid	Methyl and propyl hydroxybenzoate	• 2 :1, 0.1% mixture
Alcohol Substitutes	Chlorobutanol	• 0.5% • pH 5-6.
	Phenylethanol	• 0.5%

Chapter 29

Ophthalmic Preparations

Dosage Form

-Dr. Lokesh Adhikari, Dr. Ajay Semalty
Department of Pharmaceutical Sciences,
H.N.B Garhwal University (A Central University)
Srinagar Garhwal-246174

Lesson Plan

Manufacturing of ophthalmic preparations

Packaging of ophthalmic preparations

Preparation of eye drops involves four steps

- Preparation of the solution
- Clarification of the solution
- Filling into the container
- Sterilization of the formulation

Manufacturing of Ophthalmic Preparations

- Clarification
- Filtration through a membrane filter

Soluble Ocular Inserts

The system softens in sec after introduction into the upper conjunctival sac which gradually dissolves within one hour, while releasing the drug.

Advantage: being entirely soluble so that they do not need to be removed from their site of application.

Ophthalmic Implants

Implants are used to extend the release of drugs in ocular fluids and tissues, particularly in the posterior segment. Classified into two categories based on their degradation properties: (1) biodegradable and (2) nonbiodegradable. Implants can be solids, semisolids or particulate-based delivery systems.

Chapter 30

Ophthalmic Preparations

Dosage Form

-Dr. Lokesh Adhikari, Dr. Ajay Semalty
Department of Pharmaceutical Sciences,
H.N.B Garhwal University (A Central University)
Srinagar Garhwal-246174

Lesson Plan

Evaluation of ophthalmic preparations

Types of evaluation tests for ophthalmic products-

There are two major types of tests for ophthalmic products. Few formulation based specific tests are also mentioned in the monograph of different formulations Figure 2.

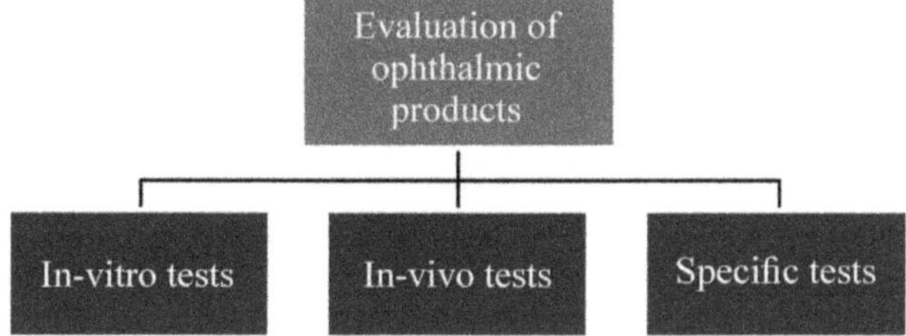

Figure 2 Types of evaluation methods of ophthalmic preparations.

In vitro **Tests include**

- Sterility
- pH
- Clarity of solutions
- Visual assessment
- Size of the particles,
- Tonicity/osmolarity
- Viscosity

- Amount of substance
- Amount of preservative
- Stability
- In vitro release

***In vivo* tests**

- Draize eye test
- *In vivo* release

Specified (Other) tests include-

- Analysis of ions (for contact lenses)
- Oxygen permeability (for contact lenses)
- Determination of encapsulation efficiency (for multicompartment drug delivery systems and emulsions)

Sterility Examination

- Thioglycolate medium
- Medium with hydrolysate of casein and soy

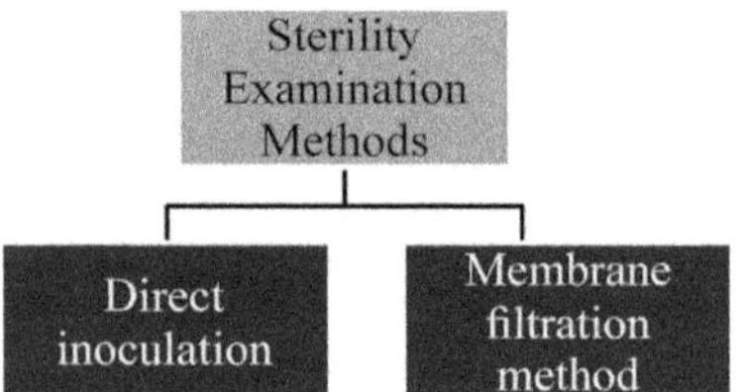

Figure 3 Sterility examination methods.

Clarity Testing (Visual Method)

Clarity Testing by UV Transmittance

Stability Analysis

In Vitro Drug Release Analysis

In vitro drug release analysis is done through varios methods (**Figure *4***).

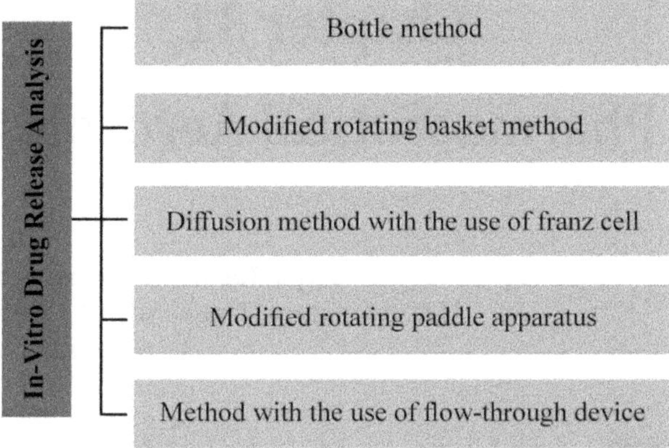

Figure 4 Types of *in-vitro* analysis methods.

Chapter 31

Pharmaceutical Aerosols I

-Dr. Ajay Semalty
Department of Pharmaceutical Sciences,
H.N.B Garhwal University (A Central University)
Srinagar Garhwal-246174

Lesson Plan

- Aerosol: Definition
- Advantages,
- Limitation
- Uses of aerosol

Definition

The word aerosol means: "Aero= air, Sol = solution". Aerosol is a "Pressurized package".

"Aerosols are a suspension of small solid particles or droplets suspended in a gas or vapor."

"Aerosols are the products that depend on the power of a compressed or liquefied gas to expel the contents from the container."

"Pharmaceutical aerosols are products that are packaged under pressure and contain therapeutically active ingredients that are released upon activation of an appropriate valve system. They are intended for topical application to the skin as well as local application into the nose (nasal aerosols), mouth (lingual aerosols), or lungs (inhalation aerosols). These products may be fitted with valves enabling either continuous or metered-dose delivery; hence, the terms "[DRUG] Metered Topical Aerosols," "[DRUG] Metered Nasal Aerosols," etc." [USP definition]

Advantages

- **Easy administration**
- **Portable**
- **Minimum dose required**

- **Minimize adverse and side effects**
- **Direct delivery of drug to target area**
- **Better spreading due to spray/ mist type of delivery**
- **Quick onset of action**
- **Lower irritation**
- **Dose to dose to maintenance of sterility**
- **Better protection for oxygen and or moisture sensitive drugs**
- **Avoid first pass metabolism**

Limitation

- Not Cost effective (diclo spray v/s diclo tablet)
- Proper use needed to get the effect. (prone to incorrect use)
- Limited safety hazard
    - Flammable
    - Pressurized
    - Inadvertent inhalation
- Ancillary devices are required with an MDI (like holding chambers and spacers)

Uses of Pharmaceutical Aerosols

- Oral and Lingual Aerosols
- Topical Aerosols
- Vaginal Aerosols
- Rectal Aerosols
- Nasal Aerosols
- Ocular Aerosols
- Respiratory Aerosols

Chapter 32

Pharmaceutical Aerosols II

(Components of Aerosols)

-Dr. Ajay Semalty
Department of Pharmaceutical Sciences,
H.N.B Garhwal University (A Central University)
Srinagar Garhwal-246174

Lesson Plan

- Propellants : concept
- Classification
- Nomenclature
- Types
- Advantages / limitations
- Uses
- Container (material)

Components of Aerosols

An aerosol system has following major components.

- Propellant
- Container
- Valve and actuator
- Product concentrate

Propellant

- Propellant is the heart of an aerosol package.
- It is develops the pressure within the container to expel the product when the valve is opened. It helps in the atomization or form production of the product.

- Propellants provide the driving force to expel product from its container.

"It is a dispersion medium as well as the driving force provider for dispensing/discharge of product from container."

Types of Propellants

The propellants are of two types

I. **Liquefied gas** e.g.Hydrocarbons (HCs), Chloroflurocarbon (CFCs), Hydrofluoroalkanes (HFA)

II. **Compressed Gas Propellants:** CO_2 ,N_2 , N_2O

Types of LG Propellants

- Hydrocarbons (HCs),
- Chloroflurocarbon (CFCs),
- Hydrofluoroalkanes (HFA)

Hydrocarbons (HCs)

- **n-Butane,**
- **isobutane,**
- **propane**

Chlorofluorcarbons (CFCs)

- CFC-11 (CCl3F),
- CFC-12 (CCl2F2),
- CFC-114 (C2Cl2 F4)**

Hydrofluoroalkanes (HFAs)

- 1,1,1,2,3,3,3-Hepta-fluoropropane (HFA-227)
- 1,1,1,2 – Tetra-fluoroethane (HFA-134a)

Nomenclature of CFCs

The **A**merican **S**ociety of **H**eating, **R**efrigerating, and **A**ir Conditioning **E**ngineers (**ASHRAE**) has created a standard for a Halocarbon Numbering System. In this system the CFCs and HFAs are designated with 3 digits followed by an alphabet if required.

- I digit = No. of Carbon atoms −1;
- II digit = No. of hydrogen atoms + 1;
- III digit = No. of Fluorine atoms;

- All remaining atoms are assumed to be chlorine atoms. An initial zero (indicating a one-carbon compound) is omitted.
- Most symmetric structure is designated by numbers only and as the asymmetry increases a, b, c, etc. follow the number

Container

Aerosol's container must withstand pressure as high as 140 to 180 psig (pounds per sq. inch gauge) at 130 ^{0}F (Table 17). Containers may be made of followings:

A. Metals
   1. Tinplated steel
   2. Tin free steel
   3. Aluminum
   4. Stainless steel

B. Glass
   1. Uncoated glass
   2. Plastic coated glass

Table 17 Maximum tolerable pressure by different aerosol containers

Container	Maximum pressure (psig)	Temperature ^{0}F
Tinplated steel	140 to 180	130
Aluminum	Up to 180	130
Stainless steel	Up to 180	130
Plastic	Less than 25	70
Uncoated glass	Less than 18	70
Coated glass	Less than 25	70

Chapter 33

Pharmaceutical Aerosols III
(Components and Systems of Aerosols)

-Dr. Ajay Semalty
Department of Pharmaceutical Sciences,
H.N.B Garhwal University (A Central University)
Srinagar Garhwal-246174

Lesson Plan

- Valve and actuators
- Aerosol systems
- Different type of systems
- Manufacturing of aerosols

Components of Aerosols

An aerosol system has following major components.

- Propellant
- Container
- Valve and actuator
- Product concentrate

Valve

It is the vital component of an aerosol package.

Major functions of valves are as followed: (RPG)

- **R**egulating the flow of product out of the container
- **P**roviding discharge of desired amount (when needed) and preventing loss when not in use.
- **G**overning the characteristics of dispensed product.

Types of Valves

- *Continuous spray valve*
- *Metering valves*

Components of Valves

Generally pharmaceutical aerosol valve comprises of following components.

- **Valve Cup (Mounting Cup/ferulae)**
- **Outer Gasket**
- **Valve Housing (Valve body)**
    - *Valve Stem*
    - *Inner Gasket*
    - *Valve Spring*
    - *Dip Tube*
- **Actuator**

Classification of Aerosol Systems

- **Propellant or container based**

  Liquefied gas systems

  Compressed gas systems

  Barrier type systems
- **Dosage form based**

  Solution systems

  Suspension or Dispersion systems

  Emusion systems
- Liquified gas systems
    - **Two phase system**
    - **(space spray, surface coating spray)**
    - **Three phase system**
    - (aquasol, foam system)

Improving the stability of aerosol dispersions

- Lower the Moisture content (< 300 ppm)
- Use API's derivative
- Reduce particle size < 5 μ (50-100 μ for topical aerosols)

- Adjust density of propellants and /or suspenoid
- Use dispersing agent like surfactant with HLB < 10 *e.g.* Sorbiton trioleate (STO), Sorbiton mono-oleate (SMO), Isopropyl myristate (IPM), Minerol oil.

Chapter 34

Pharmaceutical Aerosols IV
(Inhalers and Evaluation of Aerosols)

-Dr. Ajay Semalty
Department of Pharmaceutical Sciences,
H.N.B Garhwal University (A Central University)
Srinagar Garhwal-246174

Lesson Plan

- Type of aerosol inhalers
- Evaluation of Aerosol

Types of Aerosol Inhalers

There are three common types of aerosol generators for inhaled drug delivery of drugs in treating pulmonary diseases.

- Metered dose inhalers (MDIs)
- Dry powder inhalers (DPIs)
- Small volume nebulizers (SVN)

Metering valves are the most critical component of MDIs. The metering valve crimped onto the container is the most critical component of the pMDI (pressurized metered-dose inhaler), and has a volume ranging from 25 μL to 100 μL.

Small Volume Nebulizers (SVN)

"the devices which deliver solution and suspension in the aerosol form capable to be inhaled into Lower Respiratory Tract (LRT)"

Three types of SVN

- Pneumatic jet nebulizers
- Ultrasonic nebulizers
- Mesh nebulizers

Dry Powder Inhalers (DPIs)

DPIs are very common in anti-asthmatic drug delivery.

Small portable, propellant free, breath-actuated inhalers.

Evaluation of Aerosols

- Flammability and combustibility
- Physiochemical characteristics
- Performance
- Therapeutic activity

Flammability and Combustibility

It consist of two types of tests.

- Flame Projection
- Flash Point

Physiochemical Characteristics

Physiochemical characteristics are evaluated as followed.

- Vapor pressure - Determined by pressure gauge.
- Density - Determined by hydrometer or a pycnometer.
- Moisture content - By Karl Fischer method
- Identification of propellants - By I.R spectrophotometry

Performance Test

- Aerosol valve discharge rate/Delivery Rate
- Delivered-Dose uniformity
- Net contents/ Minimum fill
- Leakage
- Particle size distribution

Delivered-Dose Uniformity for MDIs and DPIs

"Unless otherwise specified in the individual monograph, the requirements for dosage uniformity are met **if not less than 9 of the 10 doses are between 75% and 125%** of the specified target-delivered dose and **none is outside the range of 65% to 135%** of the specified target-delivered dose."

Chapter -35

Cosmetics - I

-Dr. Mona Semalty
Department of Pharmaceutical Sciences,
H.N.B Garhwal University (A Central University)
Srinagar Garhwal-246174

Lesson Plan

- Introduction to cosmetics
- Structure of skin and absorption through skin
- Classification of cosmetics
- Skin care products –creams

Cosmetic product" can be defined as "any substance or preparation intended to be put in contact with the different superficial parts of the human body (epidermis, head and body hair, nails, lips and external genital organs), or with the teeth and the mucous membrane of the mouth in order to exclusively or generally clean, perfume, modify their appearance and/or correct body odors, and/or protect or keep them in good condition“.

(Article 2 of the European regulation n°1223/2009 of November 30th 2009 concerning cosmetic products.)

Classification of Cosmetics

1. Cosmetics as per its use-
    - SKIN- Creams, Powders, lotions
    - NAIL- Nail polish
    - TEETH-Dentifrices (Tooth paste)
    - EYE- Eye creams, Eye lotion, Eye liner
    - HAIR-Shampoo, Hair remover
2. Cosmetics as per their function.
    - CURATIVE - e.g. Antiperspirant.
    - PROTECTIVE- e.g. Face powder.

- CORRECTIVE - e.g. Face powder.
- DECORATIVE - e.g. Lipsticks, Nail polish.

3. Cosmetics as per their physical nature
    - AEROSOLS- Hair perfumes
    - CAKES - Makeup compact powder
    - EMULSION - Cold & Vanishing creams.
    - PASTE- Tooth paste.
    - POWDER -Face & Tooth powder.
    - SOLUTION- Aftershave & hand Lotion.
    - OILS - Hair oils.
    - STICKS- lipsticks
4. Other way of classifying cosmetics
    - Herbal Cosmetics –Herbal oils, creams, shampoo
    - Medicated Cosmetics-toothpaste for sensitive
    - General general purpose

Chapter 36

Cosmetics II

(Sunscreen Preparations and Dentifrices)

-Dr. Mona Semalty
Department of Pharmaceutical Sciences,
H.N.B Garhwal University (A Central University)
Srinagar Garhwal-246174

Lesson Plan

- Sunscreen products
- Ideal properties of sunscreen
- Types of sunscreen agents
- SPF how to calculate SPF Factor affecting SPF
- Dentifrices
- ideal properties/advantages and disadvantages
- Formulation of dentifrices
- Evaluation of the same

Sunscreen Preparations

These preparations help to protect the skin from the harmful effects of UV rays. In cream, lotion and spray under various brand names. Applied to the face, neck and ears prior to going out in sun.

Evaluation of Sunscreen

- Photostability
- Water Resistance
- Other Evaluation parameters

Classification of dentifrices

- Regular cleansing dentifrices
- Medicated (therapeutic) dentifrices.

Structures around the Tooth

- Periodontal ligament
- Oral Mucosa
- Gums
- Bone
- Nerves and blood supply

Formulation of Toothpaste

The raw materials used for manufacture of toothpaste can be classified as

I Basic Ingredients

- Abrasives
- Detergents
- Humectants
- Binding agents
- Sweetening Agents
- Flavors
- Preservatives

II Special Ingredients

- Coloring agents
- Bleaching agents
- Lubricants
- Therapeutic ingredients

Evaluation of Dental Care Products

- Abrasiveness (visual observation)
- Particle size
- Volatile matter and moisture
- Limit test for heavy metals
- Consistency
- Cleansing action
- pH of the product
- Foaming property

Chapter 37

Cosmetics III

(Shampoo, Hair Dye and Lipstick)

-Dr. Mona Semalty
Department of Pharmaceutical Sciences,
H.N.B Garhwal University (A Central University)
Srinagar Garhwal-246174

Lesson Plan

- Introduction to shampoos, ideal requirements, raw material and evaluation of shampoo
- Introduction to lipstic
- Ideal properties of lipstick
- Raw material and formulation of lipstick and evaluation of the same

Shampoo

Classification of Shampoos

- Normal Shampoo
- Herbal Shampoo
- Anti-Dandruff Shampoo

Raw Materials used in the Preparation of Shampoo

- Cleansing agents
- Thickening agents
- Aesthetic additives
- Conditioning agents:
- Preservatives
- Pearlizers and opacifiers

Evaluation of Shampoo

n involves:

- Physical appearance and visual inspection (Clarity, Foam producing capability, Fluidity).
- Determination of pH
- Rheological evaluations:
- Dirt dispersion:
- Foaming ability and foam stability:
- Eye irritation test:
- Detergency and cleaning action
- Foam and foam stability

"Lipsticks are cosmetic products used to impart color and shine to the lips and also used to protect lips from cracking/ drying due to exposure of dust and different weather conditions."

Manufacturing Process of Lipsticks

- Color Grinding
- Melting & Mixing
- Molding
- Flaming
- Packaging

Evaluation of Finished Products

- Color control
- Determination of melting point
- Softening point
- Microbial testing
- Breaking load test
- Rancidity

Hair Dye

Classifications of hair color

- Temporary
- Semi-Permanent
- Permanent
- Lighteners and bleaches

Evaluation Parameter of Hair Dye

- Consistency
- Spreadabillity
- Color Uniformity
- pH
- Allergy test

Chapter 38

Packaging Materials Science- I (Materials)

-Dr. Ajay Semalty
Department of Pharmaceutical Sciences,
H.N.B Garhwal University (A Central University)
Srinagar Garhwal-246174

Lesson Plan

- Purpose and requirements of Pharmaceutical packaging
- Types and properties of materials for packaging
- Containers
- Closures

Pharmaceutical Packaging

Pharmaceutical packaging means the combination of components necessary to

- Contain,
- Preserve,
- Protect
- Deliver a safe, efficacious drug product,

Role of Packaging (IPIPC)

I have an acronym for the role of packaging (IPIPC).

- Identification (green strip)
- Protection
    - light
    - moisture
    - air/reactive gases
    - microbes
    - physical damage
    - adulteration

- Information
- Presentation
- Convenience

Types of Packaging Systems

- Primary package system
- Secondary or tertiary package system

Pharmaceutical Containers

"A container for a pharmapoeial article is intended to contain a drug substance or drug product with which it is, or may be in direct contact. The closure is a part of the container." IP

Types of container (IP)

"**Airtight container**. A container that is impermeable to solids, liquids and gases under ordinary conditions of handling, storage and transport. If the container is intended to be opened on more than once, it must be so designed that it remains airtight after re-closure."

"***Hermetically Sealed container***. A container that is impervious to air or any other gas under normal conditions of handling, shipment, storage and distribution, e.g. sealed glass ampoule, gas cylinder etc."

"**Light-resistant container**. A container that protects the contents from the effects of actinic light (UV) by virtue of the specific properties of the material of which it is made."

Tamper-evident container: A container fitted with a device or mechanism that reveals irreversibly whether the container has been opened.

- **Tightly-closed container.** A tightly-closed container protects the contents from contamination by extraneous liquids, solids or vapours, from loss or deterioration of the article from effervescence, deliquescence or evaporation under normal conditions of handling, shipment, storage and distribution. A tightly-closed container must be capable of being tightly reclosed after use.

Well-closed container. A well-closed container protects the contents from extraneous solids and liquids and from loss of the article under normal conditions of handling, shipment, storage and distribution.

Packaging Materials & Closures

- Glass
- Plastic
- Metals
- Paper and Board
- Rubber
- Cotton
- Adhesives and Inks
- Closures

Types of Glass

- Type I – Borosilicate Glass
- Type II – Treated Soda-Lime Glass
- Type III – Regular Soda-Lime Glass
- Type NP – General Purpose Soda-Lime Glass

Chapter 39

Packaging Materials Science- II
(Official Requirements & Stability Aspects)

-Dr. Ajay Semalty
Department of Pharmaceutical Sciences,
H.N.B Garhwal University (A Central University)
Srinagar Garhwal-246174

Lesson Plan

- Legal requirements of Pharmaceutical packaging
- Official guidelines for pharmaceutical packaging
- Stability aspects

Legal Requirements: Pharmaceutical Packaging

Now, dear learners, let us see what are the legal requirements.

The regulator sees the pack as having the following characteristics. Regulator means the regulatory body, it may be FDA, it may be CDSCO. So, different country and different regulatory agencies in general have the same concept with respect to these factors:

- Containing the product
    - Protection of the product
    - Protection of the consumer
    - Dosage control
- Carrying the label
    - Legal control of the product
    - Informing the recipient
- Contaminating the environment
    - Packaging waste
    - Ozone depletion

- Protecting the consumer
    - Child-resistant closures
    - Tamper-evidence.

There are several factors, as below

A. Increasing sophistication of the pack
B. Increasing sophistication of the product
C. Incorporating a device into the pack
D. Cost of development

Special Types of Closure

Tamper-Evident Closures

There are 11 technologies capable of satisfying the definition of tamper-evident packaging.

- Film wrappers
- Blister packs
- Bubble packs
- Heat-shrunk bands or wrappers
- Paper foil or plastic packs
- Bottles with inner mouth seals
- Tape seals
- Breakable cap-ring systems
- Sealed tubes or plastic blindend
- Heat-sealed tubes
- Sealed cartons
- Aerosol containers and all metal and composite cans

Child-Resistant Closures

These are the closures which are difficult for drug packaging to be opened by young children while allowing adults the easy access. These are designed to minimize accidents involving the drug intoxication of children.

There are the three most common reclosable child-resistant types of closure are

- The "press–turn"
- The "squeeze–turn"
- A combination lock

Stability Test Protocols

- ICH guidelines
- Compression Strength Testing
- Distribution Simulation Testing
- Package Strength Testing
- Package Integrity Testing

Chapter 40

Packaging Materials Science- III
(QC tests of Packaging Materials)

-Dr. Ajay Semalty
Department of Pharmaceutical Sciences,
H.N.B Garhwal University (A Central University)
Srinagar Garhwal-246174

Lesson Plan

QC tests for packaging materials

- Glass
- Plastic
- Metal
- Closures

QC Test for Glass

The QC test for glass are as followed (as per Pharmacopeias)

- Hydrolytic Resistance
- Powdered glass test.
- Water attack test.
- Arsenic limit test.
- Light transmission test

Hydrolytic Resistance (IP)

The test is done to determine whether the alkali leached form the surface of a container is within the specified limits or not. The amount of acid that is necessary to neutralize the released alkali from the surface is estimated in this test. The leaching of alkali is accelerated using elevated temperature for a specified time. Methyl red (acid-base titration) is used as end point indicator.

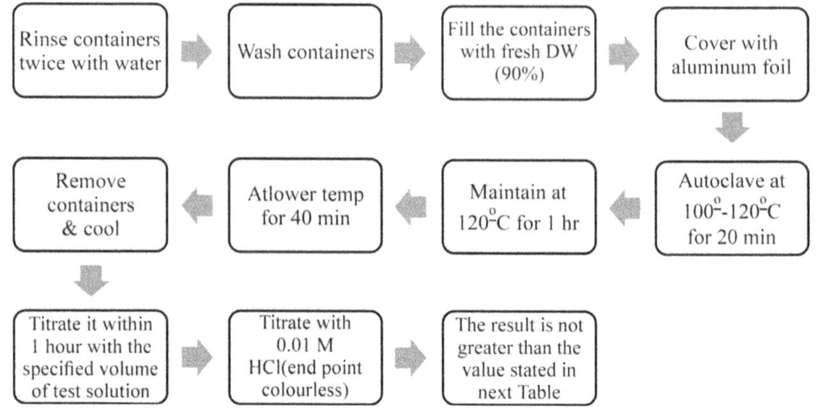

Figure 5 Test 1 -Surface glass test.

Powdered Glass Test USP (PGT)

The PGT is performed to estimate the amount of alkali leached from the powdered glass at the elevated temperatures. Leaching is assessed by acid base titration with 0.02N sulphuric acid using methyl red as an indicator.

- Step-1: Preparation of glass specimen:
- Step-2:Washing the specimen:

The acceptance criteria or the limit of Volume of 0.02N sulphuric acid (mL) is mentioned in Table 18.

Table 18 USP acceptance criteria PGT

Type of CONTAINER	Volume of 0.02N sulphuric acid (mL)
Type I	1.0
Type II	8.5
Type III	15.0

Water Attack Test (USP)

The test is done only for treated soda lime glass containers under the controlled humidity conditions which neutralize the surface alkali and glass become chemically more resistant

Principle: to detect the leached alkali from the surface of the container.

Brief Procedure (WAT)

- Rinse thoroughly with high purity water.
- Fill each container to 90% of its overflow capacity with water

- autoclaved at 121ºC for 60min then it is cooled and the liquid is decanted which is titrated with 0.02N sulphuric acid using methyl red as an indicator.
- The volume of sulfuric acid consumed is the measure of the amount of alkaline oxides present in the glass containers.

Table 19 USP acceptance criteria WAT

Type of CONTAINER	Volume of 0.02N sulphuric acid (mL)
Type II (100 mL or below)	0.7
Type II (About 100 mL)	0.2

QC Test for Plastic Containers

A. Non injectable preparations

- Collapsibility test
- Clarity of aqueous extract
- Non-volatile residue

B. Injectable preparations

- Collapsibility test
- Leakage (as in glass test)
- Clarity and colour of solution
- Acidity or alkalinity
- Reducing substances
- Light absorption
- Water vapor permeability
- Transparency

Quality Control of Closures

- Sterility
- Fragmentation
- Self sealability
- pH of aqueous extract
- Light absorption test
- Reducing substances
- Residue on evaporation

Glossary

1. Preformulation I (Physical Form: Crystal & Amorphous)

Active pharmaceutical in gradient (API)/ Drug: API or drug is a chemical entity with affinity to the receptors and with positive or negative or nil intrinsic activity.

Amorphous: These are the solids which do not exhibit long-range order in any of the three physical dimensions.

Cocrystals: Two or more molecules hydrogen bonded to each other are called cocrystals.

Crystals: These are the solids with the internal structure having long-range order in all three dimensions.

Polymorphism: The ability of a substance to exist as two or more crystalline phases that have different arrangements and/or conformations of the molecules in a crystalline lattice is called polymorphism.

Preformulation: All the activities of characterization of physicochemical properties of the drug under study which are important to develop a stable effective and safe dosage form.

2. Preformulation I (Polymorphism, Particle Size & Shape)

Conformational polymorphism: When a nonconformationally rigid molecule can be folded into alternative crystal structures, the polymorphism is categorized as conformational polymorphism.

Enantiotropic Polymorphs: One polymorph can be reversibly changed into another one by varying the temperature or pressure

Monotropic Polymorphs: The change between the two forms is irreversible

Packing polymorphism is the formation of different crystal lattices of conformationally rigid molecules that can be rearranged stably into different 3D structures through different intermolecular mechanisms.

Polymorphism: The ability of a substance to exist as two or more crystalline phases that have different arrangements and/or conformations of the molecules in a crystalline lattice is called polymorphism.

Pseudopolymorphism: The phenomenon whereby solvent or water is incorporated in the crystal lattice or in interstitial voids, is termed as pseudopolymorphism.

3. Preformulation I
Solubility Profile (Solubility, pH and pKa)

Intrinsic solubility (S_0): Solubility of a pure weak acid in acid and pure weak basic in alkali is calledIntrinsic solubility (So). It is the **fundamental solubility when completely unionized.**

pH is the negative logarithm of hydrogen ion concentration.

pH: quantitative measure of the acidity or basicity of aqueous or other liquid solutions. The term, widely used in chemistry, biology, and agronomy, translates the values of the concentration of the hydrogen ion—which ordinarily ranges between about 1 and 10−14 gram-equivalents per litre—into numbers between 0 and 14.

pKa is the negative logarithm of acid dissociation constant

pKa value is the pH at which acidic or basic groups attached to molecules exist as 50% ionized and 50% nonionized in aqueous solution

Solubility: The extent of solute dissolved in a unit amount of solvent in certain conditions of temperature, and pH.

4. Preformulation I
Partition Coefficient & Flow Properties

Angle of repose, or critical angle of repose, of a granular material is the steepest angle of descent or dip relative to the horizontal plane to which a material can be piled without slumping. At this angle, the material on the slope face is on the verge of sliding

Antiadherent: The antiadherents stop the powder from sticking to the equipment as the tablet is being made. Example: Talc, Magnesium stearate, Starch derivatives.

Glidant: The glidants improve the flowability of the tablet granules or powder by reducing the friction between particles, preventing formation of lumps e.g. Talc, Corn starch, Colloids silicates.

Hyperdiscriminating solvents: Solvents less polar than octanol. E.g. oleyl alcohol, nitrobenzene, chloroform, Carbon tetra chloride, cyclohexane etc.; reflect more the lipophilicity of blood brain barrier

Hypodiscriminating solvents: Solvents more polar than octanol. E.g. butanol, pentanol; reflect more the lipophilicity of buccal region

Log P, which can be defined as the ratio of the concentration of the unionized compound,at equilibrium, between organic and aqueous phases.

5. Preformulation II Hydrolysis, Oxidation, Reduction

Arrhenius equation: Arrhenius studied the effect of temperature on the reaction rate constant (k) and expressed their relationship in this equation ($K = A\, e^{-Ea/RT}$).

Humidity: Water vapor carrying capacity of air is called humidity.

Hydrolysis is a two-stage process, where a nucleophile, such as water or thehydroxyl ion adds to, for example, an acyl carbon, to form an intermediate from which the leaving group breaks away in the second stage.

Oxidation: Removal of hydrogen atoms from a carbon atom or addition of an oxygen atom to a carbon atom.

Reduction: Removal of an oxygen atom from a carbon atom or addition of hydrogen atoms to a carbon atom is called reduction.

6. Preformulation II Racemization

Constitutional isomers are isomers having atoms bonded to different atoms,

Enantiomers: "mirror images, not superimposable are called enantiomers"

Isomers: Different compounds with same molecular formulae are called isomers.

Racemization is a process that occurs when a compound undergoes a reaction in which the transformation produces an equal mixture of both possible enantiomers. When two compounds are classified as enantiomers of one another, it means they are **non-superimposable mirror images**."

Racemization: is the conversion of an **optically active** substance into an optically inactive mixture of equal amounts of the dextrorotatory and levorotatory."

Stereoisomers are isomers with different 3D arrangements.

7. Preformulation II Dissolution, Permeability and BCS

Bioavailability is defined in § 320.1 of US FDA guidelines for industry as: the rate and extent to which the active ingredient or active moiety is absorbed from a drug product and becomes available at the site of action. For drug products that are not intended to be absorbed into the bloodstream, bioavailability may be assessed by measurements intended to reflect the rate and extent to which the active ingredient or active moiety becomes available at the site of action. (US FDA)

Bioequivalence is defined in as "the absence of a significant difference in the rate and extent to which the active ingredient or active moiety in pharmaceutical equivalents or pharmaceutical alternatives becomes available at the site of drug action when administered at the same molar dose under similar conditions in an appropriately designed study." - 21CFR 320.1 (US FDA)

Bioequivalence of a drug product is achieved if its extent and rate of absorption are not statistically significantly different from those of the reference product when administered at the same molar dose. (CDSCO, India)

Biopharmaceutical Classification System (BCS), is a drug development tool that allows estimation of the contribution of three fundamental factors including dissolution, solubility and intestinal permeability, which govern the rate and extent of drug absorption from solid oral dosage forms.

Biopharmaceutics is the study of the interrelationship of the physicochemical properties of the active pharmaceutical ingredient (API), and its pharmacokinetic and pharmacodynamicbehavior.

Dissolution is defined as the process by which a known amount of drug substance goes into solution per unit of time under standardized conditions.

Dissolution testing is an official t¬est recommended by all pharmacopoeias (viz. official books) for evaluating drug release of solid and semisolid dosage forms.

Highly permeable: A drug substance is considered highly permeable when the extent of intestinal absorption is determined to be 85% or higher."

Highly soluble:"A drug substance is considered highly soluble when the highest strength is soluble in 250 mL or less of aqueous media over the pH range of 1.0-6.8."

***In Vitro In Vivo* Correlation (IVIVC):** The establishment of a rational relationship between a biological property, or a parameter derived from a biological property produced by a dosage form, and a physicochemical property or characteristic of the same dosage form. (USP)

IVIVC is a predictive mathematical model describing the relationship between an in vitro property of a dosage form and a relevant in vivo response. Generally, the in vitro property is the rate or extent of drug dissolution or release while the in vivo response is the plasma drug concentration or amount of drug absorbed. (US FDA)

8. Preformulation II
Polymerization

Bioavailability is defined in § 320.1 of US FDA guidelines for industry as: the rate and extent to which the active ingredient or active moiety is absorbed from a drug product and becomes available at the site of action. For drug products that are not intended to be absorbed into the bloodstream, bioavailability may be assessed by measurements intended to reflect the rate and extent to which the active ingredient or active moiety becomes available at the site of action. (US FDA)

Bioequivalence is defined in as "the absence of a significant difference in the rate and extent to which the active ingredient or active moiety in pharmaceutical equivalents or pharmaceutical alternatives becomes available at the site of drug action when administered at the same molar dose under similar conditions in an appropriately designed study." -21CFR 320.1 (US FDA)

Drug targeting is the ability of the drug to accumulate in the target organ or tissue selectively and quantitatively, independent of the site and methods of its administration.

Modified release dosage forms is defined as "one for which the drug release characteristics of time course and/or location are chosen to accomplish therapeutic objectives not offered by the conventional dosage forms. (USP)

9. Tablets-I
Introduction

Chewable tablets are formulated and manufactured so that they may be chewed, producing a pleasant tasting residue in the oral cavity that is easily swallowed and does not leave a bitter or unpleasant aftertaste. (USP)

Delayed-release tablets: Where the drug may be destroyed or inactivated by the gastric juice or where it may irritate the gastric mucosa, the use of "enteric" coatings is indicated. Such coatings are intended to delay the release of the medication until the tablet has passed through the stomach. The term "delayed-release" is used for Pharmacopeial purposes, and the individual monographs include tests and specifications for Drug release. (USP)

Dispersible tablets are uncoated or film-coated tablets that produce a uniform dispersion in water and may contain permitted flavouring and sweetening agents. (IP)

Effervescent tablets are uncoated tablets generally containing acidic substances and either carbonates or bicarbonates which react rapidly in the presence of water to release carbon dioxide. They are intended to be dissolved or dispersed in water before administration. (IP)

Enteric-coated tablets (Gastro-resistant tablets) are delayed release tablets that are intended to resist the gastric fluid but to release their active ingredient(s) in the intestinal fluid. (IP)

Extended-release tablets are formulated in such manner as to make the contained medicament available over an extended period of time following ingestion. Expressions such as "prolonged-action," "repeat-action," and "sustained-release" have also been used to describe such dosage forms. However, the term "extended-release" is used for Pharmacopeial purposes, and requirements for Drug release typically are specified in the individual monographs. (USP)

Modified-release tablets (Sustained-release tablets) are coated or uncoated tablets containing auxiliary substances or prepared by procedures that, separately or together, are designed to modify the rate or the place at which the active ingredient is released. (IP)

Prolonged-release tablets, also known as **sustained-release tablets** or **extended-release tablets** are tablets formulated in such a manner as to make the contained active ingredient available over an extended period of time after ingestion. (IP)

Tablets are solid dosage forms containing medicinal substances with or without suitable diluents. They may be classed, according to the method of manufacture, as compressed tablets or molded tablets. (USP)

Tablets are solid dosage forms each containing a unit dose of one or more medicaments. They are intended for oral administration. (IP)

Tablets are solid, flat or biconvex dishes, unit dosage form, prepared by compressing a drug or a mixture of drugs, with or without diluents.

10. Tablets-II Manufacturing Tablets

Capping: Partial or complete separation of top or bottom crowns of a tablet from the main body of tablet

Double Impression: Faulty engraving by punches with monogram or engraving

Lamination: Separation of tablet in two or more distinct layers

Mottling: It is an unequal distribution of colors on a tablet with light and dark areas on tablet surface

Picking: Surfacematerials from a tablet that is sticking to the punch and being removed from the tablet surface is *picking.*

Sticking: It refers to tablet materials adhering to the die wall.

Tablet formation is one of the most complex process which involves the volume reduction of a blend of particles/granules (compression) followed by consolidation (into a defined solid dosage form/ tablet).

Weight Variation: Variation of tablet weight beyond the acceptable limits

11. Tablets-III
Tablet Coating

Twinning: Two tablets stick together; Most common in capsule shaped tablets

Cracking: Small, fine cracks observed on the upper, lower or side surface of tablets

Blistering: It is local detachment of film from the substrate forming a blister (due to compromised elasticity or adhesive properties.

Chipping: occurs when the film becomes dented and chipped (mostly on the edges of the tablet.)

Cratering: Volcanic-like craters appears exposing the tablet surface

Picking: "part of the film sticks to the pan resulting to some of the tablet pieces being detached from the core."

Pitting: Deformation of the core of the tablet without any visible signs of disruption of the film coating.

Blooming: "the fading or dulling of a tablet colour after a prolonged period of storage at a high temperature."

Blushing: A haziness or appearance of white specks (particles of polymer that has precipitated in the film) in the film

Mottling: Variation in the colour of tablets within a batch.

Infilling: The monogram or bisect is filled and become narrow

Orange Peel (Roughness): a rough surface like orange peel

12. Tablets-IV
QC of tablets

Dissolution is defined as the process by which a known amount of drug substance goes into solution per unit of time under standardized conditions.

Dissolution testing is an official test recommended by all pharmacopoeias for evaluating drug release of solid and semisolid dosage forms.

Disintegrant: Disintegrants, such as starch help the tablet to break down into small fragments, when it is ingested. This helps the medicine to dissolve and be taken up by the body so that it can act more quickly.

Disintegration test is a measure of the time required under a given set of conditions for a group of tablets to disintegrate into particles which will pass through a 10 mesh screen.

Friability Test: The test is designed to evaluate the ability of the tablet to withstand abrasion in packaging, handling and shipping.

13. Liquid Orals (Syrups and Elixirs)

Syrup: A viscous concentrated solution of a sugar, such as sucrose, in water or other aqueous liquid combined with other ingredients such a solution is used as a flavored vehicle for medication.

Elixirs: Elixirs are the clear, sweetened, hydro alcoholic liquid intended for oral use. This contain flavoring substances and are used either as a vehicle or for the therapeutic effect of the active medicinal agents.

Solution: A solution is a homogeneous mixture of two or more substances. A solution may exist in any phase. A solution consists of a solute and a solvent.

Co-solvents: Co-solvents are primarily liquid components that are incorporated into a formulation to enhance the solubility of poorly soluble drugs to the required levels.

Solubility: Solubility is defined as no of parts of solvents (by volume) that will dissolve one part of solute(by weight of solid or liquid). Solubility is the amount of solute that passes into solution.

pH: pH is a scale used to specify how acidic or basic a water-based solution is. Acidic solutions have a lower pH, while basic solutions have a higher pH. At room temperature (25 °C), pure water is neither acidic nor basic and has a pH of 7.

Percolation: Percolation is the process of a liquid slowly passing through a filter. It is the process of extracting the soluble constituents of a powdered drug by passage of a liquid through it.

14. Emulsion –I (Introduction Theories and Identification Tests)

Emulsion: An emulsion is a mixture of two or more liquids that are normally immiscible (unmixable or unbendable). Emulsions are part of a more general class of two-phase systems of matter called colloids.

Emulsifying Agents: Emulsifying Agent is a substance used to enable oil in water or water in oil to be uniformly dispersed as an emulsion. An emulsion can be defined as the dispersion of one immiscible liquid in another liquid.

O/W emulsion: An emulsion is called o/w emulsion when water is in continuous phase and oil is dispersed in it.

W/O emulsion: An emulsion is called w/o emulsion when oil is in continuous phase and water is dispersed in it.

Microemulsion: Microemulsionisa clear, thermodynamically stable, isotropic liquid mixtures of oil, water and surfactant, frequently in combination with a cosurfactant. The aqueous phase may contain salt(s) and/or other ingredients, and the "oil" may be a complex mixture of different hydrocarbons and olefins.

Viscosity theory: As per this theory and increase in viscosity of an emulsion will lead to an increase in stability.

Surfactant: Surfactants are compounds that lower the surface tension (or interfacial tension) between two liquids, between a gas and a liquid, or between a liquid and a solid. Surfactants may act as detergents, wetting agents, emulsifiers, foaming agents, and dispersants

15. Emulsion II (Formulations of Emulsions)

Creaming: As the dispersed droplets are subjected to the gravity force, they tend to move upward this is called creaming.

Sedimentation: As the dispersed droplets are subjected to the gravity force, they tend to move downward this is called creaming.

Emulsifying Agents: Emulsifying Agent is a substance used to enable oil in water or water in oil to be uniformly dispersed as an emulsion. An emulsion can be defined as the dispersion of one immiscible liquid in another liquid.

Phase inversion: When one type of emulsion change into another emulsion is called as phase inversion.

Coalescence: Coalescence is fusion of two or more droplets of the disperse phase forming one droplet.

Ostwald ripening: Ostwald ripening is the process whereby larger droplets grow at the expense of smaller ones, because of the transport of dispersed phase molecules from smaller to larger droplets through the intervening continuous phase.

Surfactants: Surfactants are compounds that lower the surface tension between two liquids, between a gas and a liquid, or between a liquid and a solid. Surfactants may act as detergents, wetting agents, emulsifiers, foaming agents, and dispersants.

16. Suspension Dosage Form

Suspension: A pharmaceutical suspension is a coarse dispersion (biphasic dosage form) in which insoluble solid particles are dispersed in a liquid medium.

Flocculating agents: A flocculating agent is a chemical that is added to liquids so as to promote the microscopically dispersed insoluble particles, in the liquids, to aggregate and form flocs.

Deflocculating agents: An agent that prevents fine soil particles in suspension from coalescing to form flocs.

Zeta potential: The potential difference existing between the surface of a solid particle immersed in a conducting liquid and the bulk of the liquid.

Suspending agents: Suspending agents increases the viscosity of the vehicle there by slowing down sedimentation. Most of these agents form thixotropic gels which are semisolid on standing but flow radially after shaking.

Sedimentation volume: Sedimentation volume of a suspension is expressed by the ratio of the equilibrium volume of the sediment,(Vu) to the total volume (Vo) of the suspension.

$$F = Vu/Vo.$$

Degree of flocculation: Degree of flocculation is the ratio of the sedimentation volume of the flocculated suspension (F) to the sedimentation volume of the deflocculated suspension (Fo).

$$\beta = F/Fo$$

$$\beta = (Vu/Vo)\ \text{Flocculated} / (Vu/Vo)\ \text{deflocculated}$$

17. Parenterals-I (Introduction: Preformulation of Parenterals)

Amorphous form: Amorphous form of a drug is when it exists in undefined structural form

Crystalline form: Crystalline form is when it exists in defined structural form

Parenterals: Parenterals are injectables dosage form administered directly into the blood circulation by the route other than elementary canal.

Polymorphism: Polymorphism is defined as when several crystal forms of a given chemical exhibits different physical properties.

Preformulation: Preformulation is the study of physicochemical and biological properties of drug and other excipients to ensure stability safety and efficacy.

Solubility is defined as the drug substance dissolved in a unit volume of liquid (solvent) to form a saturated solution under specific conditions of temperature and pH.

Sterility: Sterility is defined as freedom from viable and non-viable microorganisms.

18. Parenterals-I (Formulation of Parenterals)

Distillation: Distillation is a process of converting water from a liquid to its gaseous form (steam).

Isotonicity: An isotonic solution refers to 2 solutions having the same osmotic pressure across semi permeable membrane and phenomena is known as isotonicity.

Reverse Osmosis (RO): Reverse osmosis is the natural process of selective permeation of molecules through a semi permeable membrane separating two aqueous solutions of different concentrations is reversed.

Sterile water for injection: It is sterile, hypotonic, nonpyrogenic, and contains no bacteriostatic or antimicrobial agents.

Ultrafiltration: Ultrafiltration is a type of membrane filtration in which hydrostatic pressure forces a liquid against semi permeable membrane.

WFI: Water For Injection (WFI) is water of extra high quality without significant contamination and is most widely used solvent for parenteral preparation

19. Parenteral-I (Types of Parenteral Preparations)

Lyophilization: The emulsions consist of an oil, an emulsifier and a balanced blend of amino acids in a continuous phase of distilled water

SVP: According to USP:" an injection that is packaged in containers labelled as containing 100 ml or less".

LVP: Large volume parenterals or LVPs (sometimes called large volume injections) are aqueous solutions usually supplied in volumes of at least 100 ml with sizes of 250 ml, 500 ml, 1000 ml, 3000 ml, and 5000 ml most common.

Injectability: Injectability refers to the properties of the suspension during injection it includes factors as pressure or force required for injection, evenness of flow, aspiration qualities, and freedom from clogging.

Syringeabillity: Syringeability refers to the handling characteristics of a suspension while drawing it into and manipulating it in syringe. It includes characteristics such as ease of withdrawal from the container into the syringe, clogging and forming tendencies and accuracy of dose measurement.

Dry Powder: Dry powders are parenteral products required to be reconstituted before administration.

20. Parenteral-I (Plant layout for Parenterals)

Clean up area:

- It is Non–aseptic area
- It is Free from dust, fibres and micro-organisms and it should withstand moisture, steam and detergents.
- Washing is done in this area.

Preparation area:

- The ingredients are mixed and preparation is prepared for filling.
- Not essential that the area is aseptic.

Aseptic area:

- Filtration and filling into final containers and sealing is done.
- The entry of outside person is strictly prohibited.
- To maintain sterility, special trained persons are only allowed to enter and work.

Laminar air flow: A laminar air flow systems are used to provide an effective means for environmental control of an aseptic area. They are of two type:-

(a) Vertical air flow

(b) Horizontal air flow.

HEPA filter: HEPA filters are high efficiency particulate air filters. These are highly efficient and can remove 99.7% of all particles 0.3μmor larger.

Validation: It is the process of establishing documentary evidence demonstrating that a procedure, process, or activity carried out in testing and then production maintains the desired level of compliance at all stages.

21. Parenteral II (Pyrogens and Pyrogenicity)

Endotoxins: Endotoxins are part of the outer membrane of the cell wall of Gram-negative bacteria.

Gram positive bacteria: Gram-positive bacteria take up the crystal violet stain used in the test, and then appear to be purple-coloured when seen through a microscope. Ex. Streptococcus and Clostridium.

Gram negative bacteria: *Gram-negative bacteria* are bacteria that do not retain the crystal violet stain used in the gram-staining method of bacterial differentiation. Ex.Vibrio and Salmonala.

Lipoplysaccharide: Lipopolysaccharides, also known as lipoglycans and endotoxins, are large molecules consisting of a lipid and a polysaccharide composed of O-antigen, outer core and inner core joined by a covalent bond; they are found in the outer membrane of Gram-negative bacteria.

Macrophages: Macrophages induces the production of cytokines (e.g. IL-1.IL-6,IL-8,etc.), prostaglandine and leukotriene, which are powerful inflammation- and septic shock- causing mediators, and are required for the adaptive immune response.

Pyrogen: Pyrogens are metabolic products of microorganisms. A substance, typically produced by a bacterium, which produces fever when introduced or released into the blood.

22. Parenteral II (Sterility Test and Sterilization)

Sterilization: Sterilization refers to any process that eliminates, removes, kills, or deactivates all forms of life and other biological agents.

D-Value: The D-value or decimal reduction time is the time required, at a given condition or set of conditions, to achieve a log reduction, that is, to kill 90% of relevant microorganisms.

F-Value: The *value* of *F* at a certain temperature is the minimum time that the organisms, present in or on an item, have to be exposed to a hostile environment to assure sterility of the item.

Z-Value: Z-value is a term used in microbial thermal death time calculations. It is the number of degrees the temperature has to be increased to achieve a tenfold (i.e. 1 $\log_{10}$) reduction in the D-value.

Liquid sterilization: Liquid sterilization is one of the methods for sterilization when low temperature is involved. Liquid sterilization is done by using Para acetic acidand Hydrogen peroxide.

Radiation Sterilization: Radiation Sterilization employs Gamma and UV light the major target for these sterilization is microbial DNA. Radiation Sterilization is done by using radiation.

23. Capsule 1

Capsules are solid dosage forms in which the drug or a mixture of drugs is enclosed in Hard Gelatin Capsule Shells, in soft, soluble shells of gelatin, or in hard or soft shells of any other suitable material, of various shapes and capacities. They usually contain a single dose of active ingredient(s) and are intended for oral administration. (IP)

Capsules are solid dosage forms in which the drug is enclosed within either a hard or soft soluble container or "shell." The shells are usually formed from gelatin; however, they also may be made from starch or other suitable substances. (USP)

Conventional-release dosage forms: Conventional-release dosage forms are preparations showing a release of the active substance(s) which is not deliberately modified by a special formulation design and/ or manufacturing method. In the case of a solid dosage form, the dissolution profile of the active substance depends essentially on its intrinsic properties. Equivalent term: immediate-release dosage form.

Dosage Form (DF) is defined as the physical form of a dose of a chemical compound used as a drug or medication intended for administration or consumption. Common dosage forms include pill, tablet, or capsule, drink or syrup, aerosol or inhaler, liquid injection, pure powder or solid crystal (e.g., via oral ingestion or freebase smoking), and natural or herbal form such as plant or food of sorts, among many others. The route of administration (ROA) for drug delivery is dependent on the dosage form of the substance.

Gelatin is a nearly transparent, faintly yellow, odorless, and almost tasteless glutinous substance obtained by boiling in water the ligaments, bones, skin, etc., of animals, and forming the basis of jellies, glues, and the like.

Modified-release dosage forms: Modified-release dosage forms are preparations where the rate and/or place of release of the active substance(s)is different from that of a conventional-release dosage form administered by the same route. This deliberate modification is achieved by a special formulation design and/or manufacturing method. Modified-release dosage forms include prolonged-release, delayed-release and pulsatile-release dosage forms.

24. Capsule II

Amorphous form: Amorphous form of a drug is when it exists in undefined structural form.

Crystalline form: Crystalline form is when it exists in defined structural form.

Humidity: Water vapor carrying capacity of air is called humidity.

Capsules are solid dosage forms in which the drug or a mixture of drugs is enclosed in Hard Gelatin Capsule Shells, in soft, soluble shells of gelatin, or in hard or soft shells of any other suitable material, of various shapes and capacities. They usually contain a single dose of active ingredient(s) and are intended for oral administration. (IP)

Excessive heat means any temperature above 40°C (104°F). (USP)

Freezer indicates a place in which the temperature is maintained thermostatically between -25° and -10° (-13° and 14°F). (USP)

Binder or binding agent is any material or substance that holds or draws other materials together to form a cohesive whole.

Lubricant is a substance, usually organic, introduced to reduce friction between surfaces in mutual contact.

25. Capsule III

Soft Gelatin Capsule: soft gelatin capsules, also called soft gels, are thicker than hard gelatin capsules and are sometimes the gelatine is plasticized by adding glycerin or sorbitol. The thickness of the gelatin is chosen by the manufacturer according to the requirements of the encased material and the environmental conditions outside the capsule.

Modified Release Capsule: In the case of capsules, the capsule body may be coated with a material through which the drug diffuses. Or it may be a slowly dissolving coat that slowly releases the drug over time.

Active substance: Equivalent terms: active ingredient, drug substance, medicinal substance, active pharmaceutical ingredient.

Vehicle: A vehicle is the carrier, composed of one or more excipients, for the active substance(s) in a liquid preparation.

Basis: A basis is the carrier, composed of one or more excipients, for the active substance(s) in semi-solid and solid preparations.

Conventional-release dosage forms: Conventional-release dosage forms are preparations

showing a release of the active substance(s) which is not deliberately modified by a special formulation design and/or manufacturing method. In the case of a solid dosage form, the dissolution profile of the active substance depends essentially on its intrinsic properties. Equivalent term: immediate-release dosage form.

Modified-release dosage forms: Modified-release dosage forms are preparations where the rate and/or place of release of the active substance(s)is different from that of a conventional-release dosage form administered by the same route. This deliberate modification is achieved by a special formulation design and/or manufacturing method. Modified-release dosage forms include prolonged-release, delayed-release and pulsatile-release dosage forms.

26. Capsule IV

Biopharmaceutics is the study of the interrelationship of the physic-chemical properties of the active pharmaceutical ingredient (API), and its pharmacokinetic and pharmacodynamic behavior.

Dissolution is defined as the process by which a known amount of drug substance goes into solution per unit of time under standardized conditions.

Dissolution testing is an official test recommended by all pharmacopoeias (viz. official

books) for evaluating drug release of solid and semisolid dosage forms.

Modified Release Capsule: In the case of capsules, the capsule body may be coated with a material through which the drug diffuses. Or it may be a slowly dissolving coat that slowly releases the drug over time.

Modified-release dosage forms: Modified-release dosage forms are preparations where the rate and/or place of release of the active substance(s)is different from that of a conventional-release dosage form administered by the same route. This deliberate modification is achieved by a special formulation design and/or manufacturing method. Modified-release dosage forms include prolonged-release, delayed-release and pulsatile-release dosage forms.

pH: quantitative measure of the acidity or basicity of aqueous or other liquid solutions. The term, widely used in chemistry, biology ,and agronomy, translates the values of the concentration of the hydrogen ion—which ordinarily ranges between about 1 and 10 −14gram-equivalents per litre—into numbers between 0 and 14.

Shelf life is the period of time after manufacturing in which the active pharmaceutical ingredient is assured to meet applicable standards of identity, strength, quality, and purity. (USP)

27. Pellets & Pelletization

Containers: Eye Drops should be packed in tamper-evident containers. Containers should be made from materials that donot cause deterioration of the preparation as a result of diffusion into or across the material of the container or by yielding foreign substances to the preparation.

Spheronization is usually associated with spherical units formed by a special process where extrudates or agglomerates are rounded as they tumble on a rotating frictional base plate.

Extrusion consists in applying pressure to a wet mass until it passes through the calibrated openings of a screen or die plate of the extruder and further shaped into small extrudate segments. The extrudates must have enough plasticity in order to deform, but an excessive plasticity may lead to extrudates which stick to each other.

Solution & Suspension Layering of a drug on a 'starter seed' material (usually, a coarse crystal or nonpareil) can produce pellets that are uniform in size distribution and generally possess very good surface morphology. These characteristics are especially desirable when pellets will be coated for the purpose of achieving a controlled release.

Cryopelletization is a process whereby droplets of a liquid formulation are converted into solid spherical particles or pellets by using liquid nitrogen as the fixing medium at -160°C.

28. Ophthalmic Preparations Introduction, Absorption through Eye, Formulation Considerations

Eye Drops can be defined as a solution or suspension in Purified Water does not preclude the inclusion of suitable additional substances where necessary for the purposes referred to above under the requirements of the European Pharmacopoeia. However, if buffering agents are used in preparations intended for use in surgical procedures great care should be taken to ensure that the nature and concentration of the chosen agent are suitable.**(BP)**

Eye Drops are sterile, aqueous or oily solutions or suspensions of one or more medicaments intended for instillation into the conjunctival sac. They may contain suitable auxiliary substances such as buffers, stabilising agents, solubilizing agents and agents to adjust the tonicity or viscosity of the preparation.**(IP)**

Containers: Eye Drops should be packed in tamper-evident containers. Containers should be made from materials that do not cause deterioration of the preparation as a result of diffusion into or across the material of the container or by yielding foreign substances to the preparation.

Particle size: This test is applicable only to Eye Drops that are suspensions. Introduce a suitable volume of the Eye Drops into a counting cell or onto a microscope slide, as appropriate. Scan under a microscope an area corresponding to 10 µg of the solid phase. Scan at least 50 representative fields. Not more than 20 particles have a maximum dimension greater than25 µm, not more than 10 particles have a maximum dimensiongreater than 50 µm and none has a maximum dimension greaterthan 100 µm.**(IP)**

Labelling. The label states (1) the names and concentrations in percentages, or weight or volume per ml, of the active ingredients; (2) the names and concentrations of any added antimicrobial preservative; (3) that, for multiple application containers, the contents should not be used for more than 1month after opening the container; (4) that, for multiple application containers, care should be taken to avoid contamination of the contents during use; (5) that the preparation is NOTFORINJECTION; (6) the conditions under which the preparation should be stored.**(IP)**

29. Ophthalmic Preparations Dosage Form

Containers: Eye Drops should be packed in tamper-evident containers. Containers should be made from materials that do not cause deterioration of the preparation as a result of diffusion into or across the material of the container or by yielding foreign substances to the preparation.

Storage: Store in sterile containers sealed so as to protect from micro-organisms.**(IP)**

Storage: Eye Drops are supplied in tamper-evident containers. The compatibility of plastic or rubber components should be confirmed before use. Containers for multi-dose Eye Drops are fitted with an integral dropper or with a sterile screwcap of suitable materials incorporating a dropper and rubber or plastic teat. Alternatively, such a cap assembly is supplied, sterilised, separately.**(BP)**

Preparation: In preparing Eye Ointments in tropical or subtropical countries where the prevailing high temperatures otherwise make the basis too soft for convenient use, the proportions of Yellow Soft Paraffin and Liquid Paraffin specified in the individual monograph may be varied, or Hard Paraffin may be added but the proportions of active ingredients must not be changed.

Multidose preparations are supplied in containers that allow successive drops of the preparation to be administered. The containers contain at most 10 ml of the preparation, unless otherwise justified and authorised.

Eye lotions are sterile aqueous solutions intended for use in rinsing or bathing the eye or for impregnating eye dressings.

30. Ophthalmic Preparations Dosage Form

Containers: Eye Drops should be packed in tamper-evident containers. Containers should be made from materials that donot cause deterioration of the preparation as a result of diffusion into or across the material of the container or by yielding foreign substances to the preparation.

Uniformity of content: Unless otherwise prescribed or justified and authorised, single-dose powders for eye drops and eye lotions with a content of active substance less than 2 mg or less than 2 per cent of the total mass comply with test B. If the preparation has more than one active substance, the requirement applies only to those substances that correspond to the above condition.

Particle size: Semi-solid eye preparations containing dispersed solid particles comply with the following test: spread gently a quantity of the preparation corresponding to at least 10 µg of solid active substance as a thin layer. Scan under a microscope the whole area of the sample. For practical reasons, it is recommended that the whole sample is first scanned at a small magnification (e.g. × 50) and particles greater than 25 µm are identified. These larger particles can then be measured at a larger magnification (e.g. × 200 to × 500). For each 10µg of solid active substance, not more than 20 particles have a maximum dimension greaterthan 25 µm, and not more than 2 of these particles have a maximum dimension greater than50 µm. None of the particles has a maximum dimension greater than 90 µm.

Eye Ointments should be packed in small, sterilised collapsible tubes of metal or of suitable plastic fitted or provided with a nozzle of suitable shape to facilitate the application of the product without contamination and with acap. The content of such containers is not more than 5 g of the preparation.

Ophthalmic inserts are sterile, solid or semi-solid preparations of suitable size and shape, designed to be inserted in the conjunctival sac, to produce an ocular effect. They generally consist of a reservoir of active substance embedded in a matrix or bounded by a rate controlling membrane. The active substance, which is more or less soluble in lacrymal liquid, is released over a determined period of time.

31. Pharmaceutical Aerosols I

Pharmaceutical aerosols are products that are packaged under pressure and contain therapeutically active ingredients that are released upon activation of an appropriate valve system. They are intended for topical application to the skin as well as local application into the nose (nasal aerosols), mouth (lingual aerosols), or lungs (inhalation aerosols). These products may be fitted with valves enabling either continuous or metered-dose delivery; hence, the terms "Metered Topical Aerosols," "Metered Nasal Aerosols," etc. (USP)

Aerosols are a suspension of small solid particles or droplets suspended in a gas or vapor.

Aerosols are the products that depend on the power of a compressed or liquefied gas to expel the contents from the container.

32. Pharmaceutical Aerosols II (Components of aerosols)

Aerosols are a suspension of small solid particles or droplets suspended in a gas or vapor.

Aerosols are the products that depend on the power of a compressed or liquefied gas to expel the contents from the container.

Propellants provide the driving force to expel product from its container. Propellants provide dispersion medium.

Propellants: It is a dispersion medium as well as the driving force provider for dispensing/discharge of product from container.

33. Pharmaceutical Aerosols III (Components and Systems of Aerosols)

Actuator: It provides rapid and convenient means for releasing the contents from a pressurized container. An actuator fits onto the valve stem

Propellants: It is a dispersion medium as well as the driving force provider for dispensing/discharge of product from aerosol container.

Valve Cup (Mounting Cup/ferulae): It is used to attach the valve properly to the container. It is typically constructed from tinplated steel or aluminum.

Valve: It is the vital component of an aerosol package. It helps to deliver the drug in desired form. It determines the performance of a pressurized package.

34. Pharmaceutical Aerosols IV (Inhalers and Evaluation of Aerosols)

Dry Powder Inhalers (DPIs): Small portable, propellant free, breath-actuated inhalers.

Flame project test: In this test effect of an aerosol on the extension of an open flame after exposure of spray for 4 sec in flame is recorded.

Mesh nebulizers: It involves use of a mesh or plate with multiple apertures to produce a liquid aerosol.

Metered Dose Inhalers (MDIs): The aerosols which can deliver a dose from 25 μLto 100 μL of fixed volume range filled in metering chamber.

Propellants: It is a dispersion medium as well as the driving force provider for dispensing/discharge of product from aerosol container.

Small Volume Nebulizers (SVN): "the devices which deliver solution and suspension in the aerosol form capable to be inhaled into Lower Respiratory Tract (LRT)"

Ultrasonic nebulizers: In ultrasonic nebulizers, electricity is used to generate ultrasonic waves and it convert a liquid into aerosols which can respired/inhaled.

Valve: It is the vital component of an aerosol package. It helps to deliver the drug in desired form. It determines the performance of a pressurized package.

35. Cosmetics - I

Cold cream: Cold cream is water in oil emulsion used for cleansing and softening of the skin.

Cosmetics: Cosmetics are any substance or preparation intended to be put in contact with the different superficial part of human body or with the teeth and the mucous membrane of mouth in order to exclusively or generally clean, perfume, modify their appearance and/or correct body odors and/or protect or keep them in good condition.

Dermis: Dermis is the thick layer of living tissue below the epidermis which forms the true skin, containing blood capillaries, nerve endings, sweat glands, hair follicles, and other structures

Epidermis: Epidermis is the upper or outer layer of the two main layers of cells that make up the skin. The *epidermis* is mostly made up of flat, scale-like cells called squamous cells

Humectants: Humectants are added to prevent excessive drying out of cream. Glycerol, sorbitol and propylene glycol and sorbitol are used as humectants.

L-Ascorbic acid: It is one of the forms of vitamin C, with age and sun exposure collagen synthesis in skin decreases leading to wrinkles. It is the only anti-oxidant proven to stimulate collagen synthesis minimizing lines, scar and wrinkles.

Vanishing cream: Vanishing cream is oil in water emulsions which leaves no visible trace when rubbed on skin.

36. Cosmetics II (Sunscreen Preparations and Dentifrices)

Binding agents: Binding agents are used in toothpaste to control viscosity and maintain creamy consistency they also prevent separation of toothpaste. Hydrocolloids are used for this purpose in concentration of 1 –

2%. Other examples includes mucilages of karayagum, tragacanth and gum arabic.

Dentifrices: Dentifrices are oral health care products which are used for improving and maintain oral hygiene and it also provides cosmetic and therapeutic benefits. It can be classified as regular dentifrices and medicated (therapeutic) dentifrices.

Humectants: Humectants prevents loss of water and subsequent hardening of toothpaste when it is expose to air. It imparts significant plasticity to the product, commonly used humectant in toothpaste are glycerine, sorbitol and propylene glycol.

Organic sunscreen agents: Organic sunscreen agents reflect UV light absorption and blocks the penetration of UV radiation through the epidermis. Example; Paraminobenzoic acid (PABA), PABA esters, salicylates, butyl-methoxy-dibenzoyl-methane.

Physical sunscreen agents: Physical sunscreen agents are photo stable components of routine sunscreen products they protect the skin by reflecting and scattering UV rays. Example Titanium dioxide (TiO2) and Zinc oxide (ZnO).

Sunscreen: Sunscreen preparation protects the skin from harmful effects of UV rays. These preparations are applied to skin prior to going out in sun.

Surfactant: Surfactant is a substance which tends to reduce the surface tension of a liquid in which it is dissolved. Example; Sodium lauryl sulphate, magnesium lauryl sulphate, sodium lauryl sacosinate

37. Cosmetics III Shampoo, Hair Dye and Lipstick

Shampoo: Shampoo are cosmetic preparation used for cleansing of hairs. It includes dust, oil, sebum and other debris it gives clean, soft, shining and strong hair.

Lipstick: Lipstick is a cosmetic product containing pigments, oils, waxes, and emollients that apply color, texture, and protection to the lips.

Surfactants: Surfactants are compounds that lower the surface tension (or interfacial tension) between two liquids, between a gas and a liquid, or between a liquid and a solid. Surfactants may act as detergents, wetting agents, emulsifiers, foaming agents, and dispersants.

Preservative: A preservative is a substance or a chemical that is added to products such as food, beverages, pharmaceutical drugs, paints, biological samples, cosmetics, wood, and many other products to prevent decomposition by microbial growth or by undesirable chemical changes.

Anti-oxidant: Antioxidants are compounds that inhibit oxidation. Oxidation is a chemical reaction that can produce free radicals, thereby leading to chain reactions that may damage the cells of organisms.

Rancidity: Rancidity is the complete or incomplete oxidation or hydrolysis of fats and oils when exposed to air, light, or moisture or by bacterial action, resulting in unpleasant taste and odour.

38. Packaging Materials Science- I Materials

Airtight container. A container that is impermeable to solids, liquids and gases under ordinary conditions of handling, storage and transport. If the container is intended to be opened on more than once, it must be so designed that it remains airtight after re-closure

Blooming/ weathering: When glassware is stored for several months, especially in a damp atmosphere or with extreme temperature variations, the wetting of the surface by condensed moisture (condensation) results in salts being dissolved out of the glass. This is called "blooming" or "weathering" & it gives the appearance of fine crystals on the glass.

Hermetically Sealed container. A container that is impervious to air or any other gas under normal conditions of handling, shipment, storage and distribution, e.g. sealed glass ampoule, gas cylinder etc.

Light-resistant container. A container that protects the contents from the effects of actinic light (UV) by virtue of the specific properties of the material of which it is made

Packaging is the science, art and technology of enclosing or protecting products for distribution, storage, sale, and use.

Tamper-evident container: A container fitted with a device or mechanism that reveals irreversibly whether the container has been opened.

Tightly-closed container. A tightly-closed container protects the contents from contamination by extraneous liquids, solids or vapours, from loss or deterioration of the article from effervescence, deliquescence or evaporation under normal conditions of handling, shipment, storage and distribution. A tightly-closed container must be capable of being tightly reclosed after use.

Well-closed container. A well-closed container protects the contents from extraneous solids and liquids and from loss of the article under normal conditions of handling, shipment, storage and distribution.

39. Packaging Materials Science- II Official Requirements & Stability Aspects

Child-resistant closures: These are the closures which are difficult for drug packaging to be opened by young children while allowing adults the easy access. These are designed to minimize accidents involving the drug intoxication of children.

Packaging is the science, art and technology of enclosing or protecting products for distribution, storage, sale, and use.

Shelf life is the period of time after manufacturing in which the active pharmaceutical ingredient is assured to meet applicable standards of identity, strength, quality, and purity. (USP)

Stability Test: The set of tests which aim to provide reasonable assurance that the products will remain at an acceptable level of fitness/quality throughout the period during which they are in market place available for supply to the patients and will be fit for their consumption until the patient uses the last unit of the product.

Stability Testing Protocols: The protocol for stability testing is a pre-requisite for starting stability testing and is necessarily a written document that describes the key components of a regulated and well-controlled stability study. In addition, the protocol can depend on whether the drug is new or is already in the market.

Tamper-evident closures: Having an indicator or barrier to entry which, if breached or missing, can reasonably be expected to provide visible evidence to consumers that tampering has occurred.- USFDA

Tamper-evident container: A container fitted with a device or mechanism that reveals irreversibly whether the container has been opened.

40. Packaging Materials Science- III QC Tests of Packaging Materials

Hydrolytic resistance (IP: This QC glass test is done to determine whether the alkali leached form the surface of a container is within the specified limits or not.

Light Transmission Test: It is a QC glass test which determines maximum % light transmission at any wavelength between 290-450nm.

Powdered Glass Test (PGT): The PGT is performed to estimate the amount of alkali leached from the powdered glass at the elevated temperatures. Leaching is assessed by acid base titration with 0.02N sulphuric acid using methyl red as an indicator.

Water Attack test (USP): The test is done only for treated soda lime glass containers under the controlled humidity conditions which neutralize the surface alkali and glass become chemically more resistant

Frequently Asked Question

1. Preformulation I (Physical Form: Crystal & Amorphous)

Q 1. What is API?

A 1. API stands for Active pharmaceutical in gradient or Drug. It is a chemical entity with affinity to the receptors and with positive or negative or nil intrinsic activity.

Q 2. What are the three key words indicating the aim of preformulation?

A 2. Stable, Effective and Safe

Q 3. Which form shows higher solubility: amorphous or crystal?

A 3. Amorphous form.

Q 4. The type of crystal with three equal axes each at right angle is called?

A 4. Simple Cubic Crystal

Q 5. DSC stands for?

A 5. Differential Scanning Calorimetry.

Q 6. What can be analysed by SEM in general?

A 6. Particle Size and Shape.

Q 7. Bragg's Law is the principle of ?

A 7. X ray diffraction.

2. Preformulation I (Polymorphism, Particle Size & Shape)

Q 1. What is polymorphism?

A 1. The ability of a substance to exist as two or more crystalline phases that have different arrangements and/or conformations of the molecules in a crystalline lattice is called polymorphism.

Q 2. What are monotropes?

A 2. The polymorphic forms in which the change between the two forms is irreversible

Q 3. Why does hydrate show slower dissolution rates?

A 3. The hydrogen bonding networkcontributes to the coherence of the crystal and hence hydrates usually show slower dissolution rates compared to the corresponding anhydrates.

Q 4. What are the commonly used lab methods of particle size distribution?

A 4. Sieving and optical microscopy method.

Q 5. Can we used DLS for microparticles?

A 5. No, it is used for nanoparticles only (Generally upto 8 micron only)

3. Preformulation I
Solubility Profile (Solubility, pH and pKa)

Q 1. What is solubility?

A 1. The extent of solute dissolved in a unit amount of solvent in certain conditions of temperature, and pH.

Q 2. What are the most common salts in pharma industry?

A 2. Sodium salt

Q 3. What is intrinsic solubility?

A 3. Solubility of a pure weak acid in acid and pure weak basic in alkali is calledIntrinsic solubility.

Q 4. What should be the ideal pH range of injections?

A 4. Injections should be in range of pH 3-9

Q 5. What is pKa?

A 5. pKais acid dissociation constant. pKavalue is the pH at which acidic or basic groups attached to molecules exist as 50% ionized and 50% nonionized in aqueous solution.

4. Preformulation I Partition Coefficient & Flow Properties

Q 1. What is partition coefficient?

A 1. The lipophilicity of an organic compound is usually described in terms of a partition coefficient,log P, which can be defined as the ratio of the concentration of the unionized compound,at equilibrium, between organic and aqueous phases:
Log P = [Unionized API] org / [unionized API] aq

Q 2. What are hypodiscriminating solvents?

A 2. **Hypodiscriminating solvents**: Solvents more polar than octanol. E.g. butanol, pentanol; reflect more the lipophilicity of buccal region

Q 3. What are hyperdiscriminating solvents?

A 3. **Hyperdiscriminating solvents:** Solvents less polar than octanol. E.g. oleyl alcohol, nitrobenzene, chloroform, Carbon tetra chloride, cyclohexane etc.; reflect more the lipophilicity of blood brain barrier

Q 4. What is angle of repose?

A 4. The angle of repose, or critical angle of repose, of a granular material is the steepest angle of descent or dip relative to the horizontal plane to which a material can be piled without slumping. At this angle, the material on the slope face is on the verge of sliding.

Q 5. What are the parameters of predicting flow?

A 5. Bulk density, Carr's index and angle of repose can easily predict the flow.

5. Preformulation I I Hydrolysis, Oxidation, Reduction

Q 1. What is hydrolysis?

A 6. Hydrolysis is a two-stage process, where a nucleophile, such as water or thehydroxyl ion adds to, for example, an acyl carbon, to form an intermediate fromwhich the leaving group breaks away in the second stage.

Q 2. What is oxidation?

A 7. Removal of hydrogen atoms from a carbon atom; or addition of an oxygen atom to a carbon atom; Loss of electron is oxidation.

Q 3. What is reduction?

A 8. Removal of an oxygen atom from a carbon atom; or addition of hydrogen atoms to a carbon atom; or gain of electron is called reduction.

Q 4. What is angle of repose?

A 9. The angle of repose, or critical angle of repose, of a granular material is the steepest angle of descent or dip relative to the horizontal plane to which a material can be piled without slumping. At this angle, the material on the slope face is on the verge of sliding.

Q 5. What is overages?

A 10. The oxidation problem is very common with vitamins and iron supplements. In that case, a little amount (as allowed by regulatory bodies like FIP) is added extra into the formulation to achieve the desired shelf life. This is called overages.

6. Preformulation II Racemization

Q 1. What are constitutional isomers?

A 11. Constitutional isomers are isomers having atoms bonded to different atoms, while the stereoisomers are isomers with different 3D arrangements.

Q 2. What is the difference between **enantiomers and monomers**?

A 12. Enantiomers: non superimposable mirror images (all chiral) while in case of diastereomers they are not mirror images (may or may not be chiral).

Q 3. What is **Racemization** ?

A 13. “the conversion of an **optically active** substance into an optically inactive mixture of equal amounts of the dextrorotatory and levorotatory.”

Or

Racemization is a process that occurs when a compound undergoes a reaction in which the transformation produces an equal mixture of both possible enantiomers.

Q 4. What is dextrorotatory and what is levorotatory?

A 14. When plane polarized light is passed through chiral sample, it rotates the plane polarized light either into the clockwise direction or into the counter-clockwise. If the light is rotated in the right direction (clockwise) it is dextrorotatory if light is rotated into the left direction (counter-clockwise) it is called levorotatory.

Q 5. What are the two different properties of thalidomide isomers?
A 15. Thalidomide is found in the two enantiomers: R-form and S-form.R-form is basically showing the tranquilizing effect while the S-form is teratogenic one.

Q 6. What is levocetrizine?
A 16. It is the single levorotatory isomers of cetrizine.

7. Preformulation II Dissolution, Permeability and BCS

Q 1. What is dissolution?
A 1. Dissolution is defined as the process by which a known amount of drug substance goes into solution per unit of time under standardized conditions.

Q 2. Which equation helps to express dissolution?
A 2. Modified Noyes-Whitney's equation.

Q 3. What are the names of three major theories of dissolution?
A 3. Diffusion layer Model (Film Theory); Penetration or Surface Renewal Theory (Danckwerts' Model); Interfacial Barrier Model (Double Barrier Mechanism or Limited Solvation Theory)

Q 4. What is Dissolution Test Apparatus type I IP?
A 4. Paddle type.

Q 5. What Dissolution Test Apparatus type V USP and for which dosage form it is used?
A 5. Paddle over disc; for Transdermal Delivery Systems.

Q 6. What is called rapid dissolving drug product?
A 6. An IR drug product is characterized as a rapid dissolving when no less than 85% of the labeled amount of the drug substance dissolves within 30 minutes using USP Apparatus I at 100 rpm or USP Apparatus II at 50 rpm in a volume of 900 mL or less of each of the recommended media.

Q 7. What are the examples of vertical diffusion cells?
A 7. Franz diffusion cell and Keshary-Chein cell

8. Preformulation II Polymerization

Q 1. What are polymers?
A 1. Monomers joined repeatedly give rise to polymers.

Q 2. Which are the polymers which form matrix?
A 2. PVP (Polyvinyl pyrrolidone), SCMC (Sodium carboxymethyl cellulose), alginates, Xanthan gum, Xanthan gum/locust bean gum combinations and Carbopol etc.

Q 3. What are the names of the polymers which are used in enteric coating?
A 3. Cellulose acetate phthalate, hydroxypropyl methylcellulose phthalate, the copolymers of methacrylic acid and their esters and polyvinyl acetate phthalate.

Q 4. What are the names of the polymers which are used in colon targeting?
A 4. The polymers used include glassy amylase (which is mixed with ethyl cellulose), azo polymers (a group of polymer which is cleaved only by the colonic bacteria), guar gum (in the same way colonic bacteria cleave that gum), xanthan gum, Cellulose acetate and Cellulose acetate phthalates.

Q 5. HV Grade polymers are?
A 5. High viscosity grade.

9. Tablets-I Introduction

Q 1. What are tablets?
A 1. Tablets are solid, flat or biconvex dishes, unit dosage form, prepared by compressing a drug or a mixture of drugs, with or without diluents.

Q 2. Which are Modified-Release dosage forms?

A 2. According to the USP/NF the term 'modified release dosage forms' is defined as "one for which the drug release characteristics of time course and/or location are chosen to accomplish therapeutic objectives not offered by the conventional dosage forms.

Q 3. What are the Delayed Release dosage forms USP?

A 3. The dosage form that releases a discrete portion or portions of drug at a time (or times) other than promptly after administration.

Q 4. What are the glidants?

A 4. These are excipients which promote flow of granules. Glidants are used for reducing the friction between particles of tablet granules and powders during processing (mixing, moving for compression etc). e.g. Corn starch, Talc, Colloids silicates. e.g. aerosil

Q 5. What are basic steps of process of dry granulation?

A 5. Process of DG: Milling, Mixing, Slugging, Screening, Mixing with disintegrant and lubricant, Compression.

Q 6. Name any two most commonly used equipment for granulation?

A 6. Fluidized Bed Dryer (FBD), RMG (Rapid Mixer Granulator

10. Tablets-II Manufacturing Tablets

Q 1. What are basic steps of manufacturing of tablets?

A 1. Tablet manufacturing consist of three steps: MGC i.e. Mixing (drug and excipients), Granulation, Compression (Tablet punching).

Q 2. Which are the major parts of a standard tablet punching machine?

A 2. Tablet punching machine consist of following parts:

- Hopper: for temporary storing the material for compressing.
- Feed frame: for distributing the materials into the dies.
- Dies: for controlling the size and the shape of the tablet.
- Punches: for compressing the materials within the dies

Q 3. What is compaction cycle?

A 3. A typical complete tablet manufacturing cycle consists of the four steps: Die filling, fill weight adjustment, Tablet compaction and Tablet ejection (from the die).

Q 4. What are the steps of the process of compaction of a powder?

A 4. The process of compaction of a powder include following steps: Particle rearrangement, Elastic, viscoelastic and plastic deformation of particles, Fragmentation of particles, Formation of interparticulate bonds.

Q 5. Which equation express volume reduction mechanism during compression?

A 5. Volume reduction mechanism during compression is expressed by Heckel equation.

Q 6. What are major tablet defects?

A 6. Capping, lamination, picking sticking, mottling, weight variation and double impression are the major tablet defects.

11. Tablets-III Tablet Coating

Q 1. What are the major type of tablet coating?

A 1. Tablet coating may be of three major types. Sugar coating, Film coating and Compression coating

Q 2. What are the basic steps for sugar-coating process?

A 2. Steps for sugar-coating process are Sealing/waterproofing (of the tablet cores), Subcoating, Smoothing, Coloring, Polishing and Printing

Q 3. What is Shellac?

A 3. It is a sealing agent

Q 4. What is Film Coating?

A 4. Film coating is the deposition, usually by spraying method, of a thin uniform film of a polymer formulation surrounding a tablet

Q 5. Which is a dry coating method?

A 5. Compression-coating is a dry coating method of tablets.

Q 6. What is picking?

A 6. When the part of the film sticks to the pan resulting to some of the tablet pieces being detached from the core, it is called picking.

Q 7. What are the causes of mottling?

A 7. The causes of mottling are: Poor mixing, uneven spray patterns of the machinery, insufficient coating, migration of soluble dyes-plasticizers and other additives during drying.

12. Tablets-IV QC of Tablets

Q 1. What are official QC test of tablets?

A 1. Official test of tablets are dissolution, Content uniformity (weight variation), Disintegration

Q 2. Why does Weight Variation test (USP/IP) is performed and when?

A 2. Weight Variation test (USP/IP)is used for assessing batch to batch uniformity. It is applicable when the tablet contains 50 mg or more of the drug.

Q 3. What is disintegration test?

A 3. The disintegration test is a measure of the time required under a given set of conditions for a group of tablets to disintegrate into particles which will pass through a 10 mesh screen.

Q 4. What is the standard Disintegration time for film coated tablet?

A 4. 30 minutes.

Q 5. What is the standard speed of Roche friabilator?

A 5. 25 rpm.

13. Liquid Orals (Syrups and Elixirs)

Q 1. What are liquid oral?

A 1. Liquid oral are the pharmaceutical dosage form in liquid form to be administered orally liquid oral include suspension emulsion elixirs and syrups.

Q 2. What percentage of sugar is used in simple syrup?

A 2. The percentage of simple syrup as per BP is 67.7% w/w and as per USP is 85% w/w.

Q 3. What is solubility?

A 3. Solubility is defined as no of parts of solvents (by volume) that will dissolve one part of solute (by weight of solid or liquid). Solubility is the amount of solute that passes into solution.

Q 4. What are co-solvents?

A 4. Co-solvents are primarily liquid components that are incorporated into a formulation to enhance the solubility of poorly soluble drugs to the required levels.

Q 5. What are elixirs?

A 5. Elixirs are the clear, sweetened, hydroalcoholic liquid intended for oral use. This contain flavoring substances and are used either as a vehicle or for the therapeutic effect of the active medicinal agents.

Q 6. What are different methods of preparation of syrup?

A 6. The different method of preparation are-

- Solution with heat.
- Agitation without heat.
- Addition of sucrose to liquid medicaments.
- Percolation.

Q 7. What is the ideal pH of oral administration?

A 7. The ideal pH of oral administration is 5-8.

14. Emulsion-I (Introduction Theories and Identification Tests)

Q 1. What is emulsion?

A 1. An emulsion is a mixture of two or more liquids that are normally immiscible (unmixable or unbendable). Emulsions are part of a more general class of two-phase systems of matter called colloids.

Q 2. What are different types of emulsion?

A 2. There are 2 types of emulsion- oil in water(o/w) and water in oil (w/o).

Q 3. What are different identification test for emulsions?

A 3. Identification tests for emulsions are:- 1) Dye Test 2) Dilution Test 3) Electrical conductivity Test 4) Fluorescence Test 5) Cobalt Chloride Test.

Q 4. What is Viscosity theory?

A 4. According to viscosity theory increase in viscosity of emulsion increase in stability of the emulsion.

Q 5. What are emulsifying agent?

A 5. Emulsifying Agent is a substance used to enable oil in water or water in oil to be uniformly dispersed as an emulsion. An emulsion can be defined as the dispersion of one immiscible liquid in another liquid.

Q 6. What are the examples of emulsifying agents?

A 6. Agar, Sodium lauryl sulfate, polymers such as spans and tweens, sodium dioctyl sulfosuccinate, etc.

15. Emulsion-II (Formulations of Emulsions)

Q 1. What is creaming and sedimentation?

A 1. As the dispersed droplets are subjected to the gravity force they tend to move upward this is called creaming and when it move downwards called sedimentation. Creaming usually happens in o/w emulsion and sedimentation usually happens in w/o emulsions.

Q 2. What are emulsifying agents?

A 2. **Emulsifying Agents:** Emulsifying Agent is a substance used to enable oil in water or water in oil to be uniformly dispersed as an emulsion. An emulsion can be defined as the dispersion of one immiscible liquid in another liquid.

Q 3. What is phase inversion?

A 3. When one type of emulsion change into another emulsion is called as phase inversion.

Q 4. What is coalescence?

A 4. Coalescence is fusion of two or more droplets of the disperse phase forming one droplet.

Q 5. What is Ostwald ripening?

A 5. Ostwald ripening is the process whereby larger droplets grow at the expense of smaller ones, because of the transport of dispersed phase molecules from smaller to larger droplets through the intervening continuous phase.

Q 6. What are surface active agents?

A 6. Surface active agents are compounds that lower the surface tension between two liquids, between a gas and a liquid, or between a liquid and a solid. Surfactants may act as detergents, wetting agents, emulsifiers, foaming agents, and dispersants.

Q 7. What are examples of surfactants?

A 7. Alkali metal and ammonium soap such as Sodium stearate and potassium oleate.

16. Suspension Dosage Form

Q 1. What is suspension ?

A 1. A pharmaceutical suspension is a coarse dispersion (biphasic dosage form) in which insoluble solid particles are dispersed in a liquid medium.

Q 2. What are flocculating agents?

A 2. A flocculating agent is a chemical that is added to liquids so as to promote the microscopically dispersed insoluble particles, in the liquids, to aggregate and form flocs.

Q 3. What are types of surfactants?

A 3. There are 4 types of surfactantsanionic surfactants, non ionicsurfactants, cationic surfactants, amphoteric surfactants.

Q 4. What is the full form of HLB?

A 4. Hydrophilic lipophilic balance.

Q 5. What are Deflocculating agents?

A 5. An agent that prevents fine soil particles in suspension from coalescing to form flocs.

Q 6. what is deflocculated system?

A 6. The solid particles exists as a separate entities the sediment form hard cake and the rate of sedimentation is low this type of system is called deflocculated system.

Q 7. What is zeta potential?

A 7. The potential difference existing between the surface of a solid particle immersed in a conducting liquid and the bulk of the liquid.

17. Parenterals- I (Introduction: Preformulation of Parenterals)

Q 1. What are parenterals?

A 1. Parenterals are injectables dosage form administered directly into the blood circulation by the route other than elementary canal.

Q 2. What is preformulation?

A 2. Preformulation is the study of physicochemical and biological properties of drug and other excipients.

Q 3. What is sterility?

A 3. Sterility is defined as freedom from viable and non-viable microorganisms.

Q 4. What are the commonly used routes of administration of parenterals?

A 4. The commonly used routes of parenteral administration are intravenous, intramuscular and sub-cutaneous route of administration.

Q 5. What is crystalline and amorphous form of a drug?

A 5. Amorphous form of a drug is when it exists in undefined structural form and crystalline form is when it exists in defined structural form.

Q 6. What is the importance of chemical modification?

A 6. Chemical modification like preparation of an ester, salt or any other form of a parent drug to increase stability, alter drug solubility, enhances deport action, avoid formulation difficulties and possibly to decrease pain on injection.

Q 7. What is polymorphism?

A 7. Polymorphism is defined as when several crystal forms of a given chemical exhibits different physical properties.

Q 8. What is solubility?

A 8. Solubility is defined as the amount of drug substance dissolved in a unit volume of liquid (solvent) to form a saturated solution under specific conditions of temperature and pH.

Q 9. What is solvates and co-crystals?

A 9. Co-crystals are crystalline material composed of two or more molecule in the same crystal lattice and solvates are chemical substances with molecule of solvent regularly incorporated into a unique molecular packing.

Q 10. What is bioavailability?
A 10. Bioavailability is defined as the rate and extent of drug that reaches to systemic circulation.

Q 11. What are important properties of parenteral dosage form?
A 11. Important properties of parenteral dosage forms are sterility, clarity, non-pyrogenicity and stability.

Q 12. What are important preformulation parameters of parenteral?
A 12. The important preformulation parameters of parenterals are solubility, pKa, pH, solid state characteristics, chemical modification of drug and polymorphism.

18. Parenterals- I (Formulation of Parenterals)

Q 1. What are different methods of preparation of Water For Injection (WFI)?
A 1. The different methods of preparation are distillation, reverse osmosis and Ultrafilteration.

Q 2. What is distillation?
A 2. Distillation is a process of converting water from a liquid to its gaseous form (steam)

Q 3. What isultrafiltration?
A 3. Ultrafiltration is a type of membrane filtration in which hydrostatic pressure forces a liquid against semi permeable membrane.

Q 4. What is isotonicity?
A 4. An isotonic solution refers to 2 solutions having the same osmotic pressure across semi permeable membrane and phenomena is known as isotonicity.

Q 5. What are different methods of adjusting tonicity?
A 5. The different methods of adjusting tonicity are Freezing point depression, Sodium chloride equivalent method, White-Vincent method and Sprowls method

Q 6. What is reverse osmosis?
A 6. Reverse osmosis is a natural process of selective permeation of molecules through a semi permeable membrane separating two aqueous solutions of different concentration in reverse. A driving force (usually between 200 and 400 psig) is required to overcome osmotic pressure and to permeate pure water through the membrane.

Q 7. What are different types of vehicle used for the production of parenterals?

A 7. The different types of vehicles used are aqueous, nonaqueous and water-miscible.

19. Parenteral- I (Types of Parenteral Preparations)

Q 1. What is SVP?

A 1. According to USP: "an injection that is packaged in containers labelled as containing 100 ml or less".

Q 2. What is LVP?

A 2. Large volume parenterals or LVPs (sometimes called large volume injections) are aqueous solutions usually supplied in volumes of at least 100 ml with sizes of 250 ml, 500 ml, 1000 ml, 3000 ml, and 5000 ml most common.

Q 3. What is different types of vehicles use for preparation of parenteral?

A 3. Two types of vehicles used for preparation of parenteral are aqueous vehicle WFI and SWFI and non aqueous vehicle Water miscible vehicle.

Q 4. What are parenteral emulsions?

A 4. The emulsions consist of an oil, an emulsifier and a balanced blend of amino acids in a continuous phase of distilled water

Q 5. What are different methods of production of dry powders?

A 5. Different methods of production of dry powders are- a. Aseptic Crystalllization and Dry powder filling b. Spray drying c. Freeze drying

Q 6. What is lyophilization?

A 6. The emulsions consist of an oil, an emulsifier and a balanced blend of amino acids in a continuous phase of distilled water

Q 7. What is syringeability and injectability?

A 7. Injectability refers to the properties of the suspension during injection it includes factors as pressure or force required for injection, evenness of flow, aspiration qualities, and freedom from clogging.

Syringeability refers to the handling characteristics of a suspension while drawing it into and manipulating it in syringe. It includes characteristics such as ease of withdrawal from the container into the syringe, clogging and forming tendencies and accuracy of dose measurement.

20. Parenterals-I (Plant layout for Parenterals)

Q 1. What are different areas for production of sterile products?
A 1. The different areas for the production of sterile products are preparation area, clean up area , quarantine area and aseptic area.

Q 2. Which method is used for the validation of HEPA filter?
A 2. Di octyl phthalate (DOP) is used for the validation of HEPA filter.

Q 3. What are different types of laminar air flow?
A 3. There are two types of laminar air flow
(a) Vertical air flow
(b) Horizontal air flow.

Q 4. What are HEPA filter?
A 4. HEPA filters are high efficiency particulate air filters. These are highly efficient and can remove 99.7% of all particles 0.3μm or larger.

Q 5. What is laminar air flow?
A 5. A laminar air flow systems are used to provide an effective means for environmental control of an aseptic area.

Q 6. Which type of operations are carried out in aseptic areas?
A 6. Filtration and filling into final containers and sealing is done in aseptic areas.
And to maintain sterility special trained persons are only allowed to enter and to work in this area.

21. Parenterals II (Pyrogens and Pyrogenicity)

Q 1. What arepyrogens?
A 1. Pyrogens are metabolic products of microorganisms.

Q 2. What are different sources of pyrogen contamination?

A 2. The different sources of pyrogen contamination are Solvent, the medicament, apparatus, and the method of storage between prepration and sterilization.

Q 3. What are different test of pyrogen detection?

A 3. Pyrogen can be detected by LAL test and rabbit test.

Q 4. How leakers test is performed?

A 4. Leakers test is performed by incubating ampoules in 1% Methylene blue solution. The color ampoules are discarded.

Q 5. What are the different test of sterility?

A 5. The different test of sterility are direct transfer method and membrane filtration method.

Q 6. What is the structure of endrotoxins?

A 6. The structure of endrotoxins consist of the O-specific chain, the Lipid A, the R-core

22. Parenteral II (Sterility Test and Sterilization)

Q 1. What is sterilization?

A 1. Sterilization refers to any process that eliminates, removes, kills, or deactivates all forms of life and other biological agents.

Q 2. What are different sterility test method?

A 2. Direct transfer and Direct inoculation.

Q 3. What are different types of media used for sterility test method?

A 3. Fluid Thioglycolate medium and soyabean- casein digest medium.

Q 4. What is D-Value?

A 4. The D-value or decimal reduction time is the time required, at a given condition or set of conditions, to achieve a log reduction, that is, to kill 90% of relevant microorganisms.

Q 5. What is F-value?

A 5. The *value* of *F* at a certain temperature is the minimum time that the organisms, present in or on an item, have to be exposed to a hostile environment to assure sterility of the item.

Q 6. What is Z-value?
A 6. Z-value is a term used in microbial thermal death time calculations. It is the number of degrees the temperature has to be increased to achieve a tenfold (i.e. 1 $\log_{10}$) reduction in the D-value.

Q 7. What is temperature used in heat sterilization?
A 7. For dry heat 162-180 °**C** and For moist heat 121-134°**C.**

Q 8. What are commonly employed gases used in sterilization?
A 8. Formaldehyde and ethylene dioxide are commonly used gases in sterilization

23. Capsule I

Q 1. What are the capsules?
A 1. Capsules are solid preparations with hard or soft shells of various shapes and capacities, usually containing a single dose of active substance(s). They are intended for oral administration.(British Pharmacopoeia)

Q 2. What are the common types of capsules?
A 2. Hard capsules; soft capsules; gastro-resistant capsules; modified-release capsules; cachets.

Q 3. How many types of gelatinis available for capsule shell production?
A 3. Two, type A and type B.

Q 4. What is the second most preferable capsule shell forming material?
A 4. Hypromellose.

Q 5. What are the steps of capsule shell formation?
A 5. Capsule shorting, capsule printing, packaging and storage

24. Capsule II

Q 1. What are the general types of filling materials for hard capsules?
A 1. Powdered solid, pellets and granules; coated or uncoated.

Q 2. Why the compression slug method is most preferable method of processing the filling material for capsules?
A 2. The weight variation is very less in case of compression slug method.

Q 3. What is Auger fill method?
A 3. It is a semi-automatic volumetric method for filling the capsules.

Q 4. How we ensure the closing of the filled capsule?
A 4. By sealing method which can be done in two ways either by heat welding or by liquid welding.

Q 5. Why reducing the particle size of filling material is not an insurance of better performance in case hard gelatin capsule?
A 5. Because reducing particle size may lead to the agglomeration of the particles.

25. Capsule III

Q 1. What are the general types of filling materials for softgel capsules?
A 1. Liquid and semisolids including fixed oils.

Q 2. For a vegan patient, what is an alternative of gelatin as it is produced from animal skin and bones?
A 2. Carrageenan, it is produced from marine sources.

Q 3. Why gelatin is the most preferred shell formulating material for capsules?
A 3. As gelatin is biocompatible, tasteless and odourless, it is preferred over other capsule shell formulating agents.

Q 4. What is encapsulation?
A 4. Encapsulation can be defined as wrapping a core material with a thin film of a covering material.

Q 5. What are the most important factors related to the gelatin used for softgel capsule formation?
A 5. Temperature, viscosity, surface activity and particle size.

26. Capsule IV

Q 1. What are the general types of materials we can evaluate to ensure the capsules quality?
A 1. We can categories the material in three categories for testing, raw material, in process material and finished material.

Q 2. What are the limits of bloom strength of gelatin solution for best capsule shell formation?
A 2. The range for bloom strength for food grade gelatin solution should be 125 Bloom to 250 Bloom.

Q 3. For which organism we test the gelatin solution under antimicrobial testing protocols?
A 3. *E. coli* and *Salmonella*; including yeasts & molds.

Q 4. If the weight of individual capsule is more than 300 mg, what will be the limit of weight deviation of individual capsule?
A 4. 7.5%.

Q 5. Which type of apparatus is used in the dissolution testing of capsules?
A 5. Basket type dissolution apparatus.

27. Pellets & Pelletization

Q 1. What are the pellets?
A 1. Pellets are small, free-flowing, spherical particulates manufactured by the agglomeration of fine powders or granules of drug substances and excipients.

Q 2. What is powder layering?
A 2. It involves the deposition of successive layers of dry powder of drug or excipients or both on preformed nuclei/cores with the help of a binding liquid.

Q 3. Enlist the optimizable properties of powder layering technique?
A 3. The rheological properties of the binding liquid, the liquid application rate, and drying air temperature.

Q 4. What are the disadvantages of Wurster process?
A 4. Clogging of the nozzles is the main disadvantage of Wurster process.

Q 5. What are the advantages of pelletization?
A 5. Pelletization is applied to improve the drug release as well as patient compliance of any formulation.

28. Ophthalmic Preparations Introduction, Absorption through Eye, Formulation Considerations

Q 1. What is the topical route?

A 1. When the drug is given by application on to the surface or the skin, the route is defined as topical route.

Q 2. An automated reflex system of eye to remove any external undesired particle is known as?

A 2. Lacrymal secretion system.

Q 3. The type of lens which is present in human eye is?

A 3. Biconvex lens.

Q 4. The colourless gelatinous fluid filled in the eye is known as?

A 4. Vitreous humor

Q 5. What are the factors affecting the drug absorption from ophthalmic route?

A 5. Gravity, lachrymation, blinking of eye, low corneal permeability, and tear turnover are factors which affects the drug absorption from ophthalmic route.

29. Ophthalmic Preparations Dosage Form

Q 1. Which type of water is used as vehicle for ophthalmic preparations?

A 1. Water for injection.

Q 2. What is the desirable pore size of the filters used to clarify the ophthalmic solutions?

A 2. Not more than 0.25 μm.

Q 3. What is the desirable pore size of the filters used to clarify the ophthalmic suspensions?

A 3. Not more than 10 μm.

Q 4. What is the desirable pore size of the filters used to clarify the ophthalmic ointments?

A 4. Not more than 25 μm.

Q 5. What is the advantage of soluble ocular implants over insoluble ocular implants?

A 5. The soluble ocular implants tend to dissolve completely to the removal of implant is not applicable which increases patient compliance.

30. Ophthalmic Preparations Dosage Form

Q 1. What are the methods we apply to assess the sterility of ophthalmic preparations?

A 1. There are two methods to assess the sterility of an ophthalmic product: 1) Direct Inoculation and 2) Membrane Filtration.

Q 2. For fatty ointment bases, during the sterility test, which vehicle is used for dilution?

A 2. Isopropyl myristate.

Q 3. If the ophthalmic preparation of solid type such as contact lens or implants than what will be the appropriate technique for clarity testing?

A 3. UV resistance method

Q 4. How we assess whether the ionic concentration equivalent to lachrymal secretion or not?

A 4. By assessing the tonicity of the preparation.

Q 5. What is the evaluation parameter of Draize test?

A 5. Blinking rubbing of eye.

31. Pharmaceutical Aerosols I

Q 1. What are aerosols?

A 1. Aerosol is a "Pressurized package". OR Aerosols are the products that depend on the power of a compressed or liquefied gas to expel the contents from the container.

Q 2. What are the major advantages of aerosols?

A 2. Low dose, quick onset of action, high bioavailability, low adverse effect, avoiding FPM are the main advantages

Q 3. What are the limitations of aerosols?
A 3. Not Cost effective, prone to incorrect use, flammable.

Q 4. What are MDIs?
A 4. Metered Dose Inhalers

Q 5. What are DPIs?
A 5. Dry Powder Inhalers

32. Pharmaceutical Aerosols II (Components of Aerosols)

Q 1. What are liquefied gas propellants?
A 1. These are materials that at room temperature and atmospheric pressure exist in the gaseous or vapor state and are capable of being liquefied at relatively low pressures or temperatures. e.g. Hydrocarbons (HCs), Chloroflurocarbon (CFCs), Hydrofluoro-alkanes (HFA)

Q 2. What ispropellant?
A 2. It is a dispersion medium as well as the driving force provider for dispensing/discharge of product from container.

Q 3. In which type of propellants the pressure remain unchanged before and after the use?
A 3. liquefied gas propellants.

Q 4. How CFC and HFA propellants are named?
A 4. The CFCs and HFAs are designated with 3 digits followed by an alphabet if required. I digit = No. of Carbon atoms − 1; II digit = No. of hydrogen atoms + 1; III digit = No. of Fluorine atoms; All remaining atoms are assumed to be chlorine atoms. An initial zero (indicating a one-carbon compound) is omitted. Most symmetric structure is designated by numbers only and as the asymmetry increases a, b, c, etc. follow the number.

Q 5. Which are the non-environment friendly propellants and why?
A 5. Chlorofluorocarbons (CFCs) due to ozone depleting effect.

Q 6. What are the alternative propellants of CFCs?
A 6. Hydrofluoroalkanes (HFAs) e.g. 1,1,1,2,3,3,3-Hepta-fluoropropane (HFA-227), 1,1,1,2 – Tetra-fluoroethane (HFA-134a).

33. Pharmaceutical Aerosols III (Components and Systems of Aerosols)

Q 1. What are major components of aerosols?
A 1. Propellant, Container, Valve and actuator, Product concentrate

Q 2. What kind of aerosols need initial high pressure?
A 2. The one with compressed gas as propellants.

Q 3. What are the methods of aerosol manufacturing?
A 3. Pressure filling and Cold filling.

Q 4. How many types are there of Liquefied gas systems?
A 4. Two. Two phase system and three phase system.

Q 5. Which type of aerosol propellants are used at 40-50 psig at 21 ^{0}C with 4-7 % concentration?
A 5. Foam or Emulsion Systems

Q 6. What is barrier type system of aerosols?
A 6. Propellants are separated from the product with some physical barrier in these systems.

34. Pharmaceutical Aerosols IV (Inhalers and Evaluation of Aerosols)

Q 1. What is the fixed volume range of metered doseinhalers?
A 1. 25 μL to 100 μL.

Q 2. What are the advantages of pMDI?
A 2. Advantages of MDIs are Portable and compact, Short treatment time, Reproducible dose emitted.

Q 3. What are SVN?
A 3. **Small Volume Nebulizers (SVN) are** the devices which deliver solution and suspension in the aerosol form capable to be inhaled into Lower Respiratory Tract (LRT).

Q 4. What are major performance test of aerosols?

A 4. Aerosol valve discharge rate/Delivery Rate; Delivered-Dose uniformity; Net contents/ Minimum fill; Leakage; Particle size distribution.

Q 5. What are the test performed for assessing flammability and combustibility?

A 5. Flam projection and flash point.

35. Cosmetics-I

Q 1. What are cosmetics?

A 1. Cosmetics are any substance or preparation intended to be put in contact with the different superficial part of human body or with the teeth and the mucous membrane of mouth in order to exclusively or generally clean, perfume, modify their appearance and/or correct body odors and/or protect or keep them in good condition.

Q 2. What is cold cream?

A 2. Cold cream is water in oil emulsion used for cleansing and softening of the skin.

Q 3. What is vanishing cream?

A 3. Vanishing cream is oil in water emulsions which leaves no visible trace when rubbed on skin.

Q 4. What are major ingredients of vanishing cream?

A 4. Major ingredients of vanishing cream are stearic acid, humectant and alkyl.

Q 5. What are different layers of skin?

A 5. Dermis, epidermis and subcutaneous fat tissues (hypodermis).

Q 6. What are different zones of epidermis?

A 6. The different zones of epidermis are Basal layer (stratum basal), spinous layer (stratum spinosum), granular layer (stratum granulosum) and cornified layer (stratum corneum).

Q 7. What is the role of L-ascorbic acid?

A 7. It is one of the forms of vitamin C, with age and sun exposure collagen synthesis in skin decreases leading to wrinkles. It is the only anti-oxidant proven to stimulate collagen synthesis minimizing lines, scar and wrinkles.

36. Cosmetics-II (Sunscreen Preparations and Dentifrices)

Q 1. What is sunscreen?

A 1. Sunscreen preparation protects the skin from harmful effects of UV rays. These preparations are applied to skin prior to going out in sun.

Q 2. What are organic sunscreen agents?

A 2. Organic sunscreen agents reflect UV light absorption and blocks the penetration of UV radiation through the epidermis. Example; Paraminobenzoic acid (PABA), PABA esters, salicylates, butyl-methoxy-dibenzoyl-methane.

Q 3. What are Dentifrices?

A 3. Dentifrices are oral health care products which are used for improving and maintain oral hygiene and it also provides cosmetic and therapeutic benefits. It can be classified as regular dentifrices and medicated (therapeutic) dentifrices.

Q 4. What are Humectants?

A 4. Humectants prevents loss of water and subsequent hardening of toothpaste when it is expose to air. It imparts significant plasticity to the product, commonly used humectant in toothpaste are glycerine, sorbitol and propylene glycol.

Q 5. What are major evaluation parameters of sunscreen products?

A 5. Major evaluation parameters of sunscreen are photostability, water resistant and sun-protectant.

Q 6. What is SPF?

A 6. SPF is defined as sun protection factor and it is calculated by using formula SPF = MED of photoprotected skin/ MED of unprotected skin.

Q 7. What are physical sunscreen agents?

A 7. Physical sunscreen agents are photo stable components of routine sunscreen products they protect the skin by reflecting and scattering UV rays. Example Titanium dioxide(TiO2) and Zinc oxide (ZnO).

Q 8. Which is the most commonly used sweetener in dental products?

A 8. Saccharin sodium.

Q 9. What are the most commonly used preservatives in toothpaste?

A 9. Most commonly used preservative in toothpaste are methylparahydroxybenzoate (0.15%) and propylparahydroxybenzoate (0.02%).

Q 10. What is Surfactant?

A 10. Surfactant is a substance which tends to reduce the surface tension of a liquid in which it is dissolved. Example; Sodium lauryl sulphate, magnesium lauryl sulphate, sodium lauryl sacosinate.

37. Cosmetics III Shampoo, Hair Dye and Lipstick

Q 1. What are shampoo's?

A 1. Shampoo are cosmetic preparation used for cleansing of hairs. It includes dust, oil, sebum and other debris it gives clean, soft, shining and strong hair.

Q 2. Which equipment is used to measure surface tension?

A 2. Stalagmometer is used to measure surface tension.

Q 3. What are lipstick?

A 3. Lipstick's are cosmetics products used to impart colour and shine to the lips and also used to protect lips from cracking / drying due to exposure of dust and different weather conditions.

Q 4. What are the raw material used for the preparation of lipstick?

A 4. The raw material used for the preparation of lipstick are waxes, oil, pigments and dies, alcohol and fragrance, preservative and anti-oxidants.

Q 5. State few example used for the pigments and dies?
A 5. The example used for the pigments and dies are
(a) Manganese violet
(b) Titanium Di-oxide
(c) D and C red no. 6.

Q 6. What are different step involved in the manufacturing of lipstick?
A 6. The different step involved in the manufacturing of lipstick are
- Color grinding
- Melting and mixing
- Moulding
- Flaming
- Packing.

Q 7. What are different types of hair color?
A 7. Different types of hair color are temporary hair color, permanenthair color, semi permanent hair color, bleaches.

38. Packaging Materials Science- I Materials

Q 1. What is the role of Packaging?
A 1. Identification, Protection (against light, moisture, air/reactive gases, microbes, physical, damage, adulteration), Information, Presentation and Convenience. **(IPIPC)**

Q 2. Which law governs the packaging guidelines in India?
A 2. D&C Act 1940 & rules 1945.

Q 3. Which is the most highly resistant glass?
A 3. Type I: Borosilicate Glass

Q 4. What is the process which prevents "weathering" of empty glass bottles?
A 4. Sulphur treatment (de-alkalizing process)

Q 5. What are pharmaceutical containers?
A 5. A container for a pharmacopoeial article is intended to contain a drug substance or drug product with which it is, or may be in direct contact. The closure is a part of the container. – IP

Q 6. What is hermetically sealed container?

A 6. A container that is impervious to air or any other gas under normal conditions of handling, shipment, storage and distribution, e.g. sealed glass ampoule, gas cylinder etc.

39. Packaging Materials Science- II Official Requirements & Stability Aspects

Q 1. What are Child-resistant closures?

A 1. Child-resistant closures are the closures which are difficult for drug packaging to be opened by young children while allowing adults the easy access. These are designed to minimize accidents involving the drug intoxication of children.

Q 2. Which law governs the packaging guidelines in India?

A 2. D&C Act 1940 & rules 1945.

Q 3. Which is the successful and cost effective strategy between new drug discovery and Novel packaging development?

A 3. Novel packaging development is the successful and cost effective strategy.

Q 4. What are the three most common reclosable child-resistant types of closure?

A 4. The "press–turn", The "squeeze–turn", A combination lock

Q 5. What is accelerated stability testing?

A 5. It is the performance assessment of product/ package at relatively high temperatures and/or humidity.

Q 6. What is ICH?

A 6. ICH stands for International Council for Harmonisation, formerly the International Conference on Harmonisation (ICH). ICH is an organization in which several countries have joined there hand together to have the uniform guidelines on various aspects of quality control test and all the guidelines.

40. Packaging Materials Science- III QC Tests of Packaging Materials

Q 1. What is the Water Attack test?

A 1. It is a QC glass test which is performed to detect the leached alkali from the surface of the container.

Q 2. Which IP test is similar with Water Attack Test USP?

A 2. Hydrolytic resistance Test.

Q 3. Which type of glass consume the least volume of acid in Powdered Glass Test?

A 3. Type I glass.

Q 4. What is the maximum number of fragments allowed in Fragmentation test ?

A 4. 15.

Q 5. What are the common QC test for Glass?

A 5. Hydrolytic Resistance, Powdered glass test, Water attack test, Arsenic limit test, Light transmission test.

Multiple Choice Based Questions for Self-Assessment

1. Preformulation I (Physical form: Crystal & Amorphous)

Q 1. Preformulation is about ensuring

(a) Stability

(b) Safety

(c) Efficacy

(d) All of the above

Q 2. Amorphous form shows

(a) Good solubility

(b) Poor solubility

(c) Long range order

(d) None of the above

Q 3. Crystal form shows this structure

(a) 3D

(b) 2D

(c) Short range

(d) None of the above

Q 4. In simple cubic crystal the faces are

(a) All unequal

(b) Two equal one unequal

(c) All equal

(d) None of the above

Q 5. Two or more molecules are hydrogen bonded to each other are called

(a) Crystals

(b) Cocrystals

(c) Polymorph

(d) All of the above.

Q 6. DSC is
(a) Differential scanning calorimetry
(b) Differential scattering calorimetry
(c) Digital scanning calorimetry
(d) None of the above

Q 7. X ray diffraction pattern indicate
(a) Solubility
(b) Crystallinity
(c) Functional groups
(d) Permeability

Q 8. SEM can analyze
(a) Shape & Size
(b) Crystallinity & solubility
(c) Complexation
(d) Flow property

Q 9. Bragg's law define the
(a) Shape
(b) Solubility
(c) Diffraction
(d) All of the above

Q 10. Crystal form show
(a) Short duration of action
(b) Rapid onset of action
(c) Short onset of action
(d) Long duration of action

Answer Key

Q. 1	d	Q. 6	a
Q. 2	a	Q. 7	b
Q. 3	a	Q. 8	a
Q. 4	c	Q. 9	c
Q. 5	b	Q. 10	d

2. Preformulation I (Polymorphism, Particle Size & Shape)

Q 1. The ability of a substance to exist as two or more crystalline phases with different arrangements of molecules in crystal lattice is

(a) Cocrystal
(b) Crystal
(c) Polymorphism
(d) None of the above

Q 2. Estrone show show many polymorphs

(a) 4
(b) 2
(c) 3
(d) None

Q 3. Polymorphs differ in

(a) Physical properties
(b) Chemical properties
(c) a & b both
(d) None of the above

Q 4. Enantiotropic polymorph can change in to another form

(a) reversibly
(b) irreversibly
(c) does not change
(d) None of the above

Q 5. The change between the two forms is irreversible in the type of polymorph

(a) Enantiotropic
(b) Monotropic
(c) a & b both
(d) None of the above

Q 6. Hydrates and solvates are called

(a) Polymorph
(b) Pseudopolymorph
(c) Cocrystals
(d) None of the above

Q 7. DLS is used for determining
(a) Particle size
(b) Crystallinity
(c) Functional groups
(d) Shape

Q 8. Coulter Counter can analyze
(a) Size distribution
(b) Crystallinity & solubility
(c) Number of functional groups
(d) Flow property

Q 9. Most advanced particle size analysis technique is
(a) SEM
(b) DLS
(c) XRPD
(d) Sieving

Q 10. Micronization increases
(a) Size
(b) Surface area
(c) Bioactivity
(d) Duration of action

Answer Key

Q. 1	c	Q. 6	b
Q. 2	c	Q. 7	a
Q. 3	a	Q. 8	a
Q. 4	a	Q. 9	b
Q. 5	b	Q. 10	b

3. Preformulation I Solubility Profile (Solubility, pH and pKa)

Q 1. Water solubility of API is needed for

(a) Formulation development
(b) Dissolution into GI fluid
(c) A & b both
(d) None of the above

Q 2. For good bioavailability, what is needed

(a) Good solubility
(b) Good permeability
(c) Good solubility poor permeability
(d) Good solubility and good permeability

Q 3. Less soluble salts are intentionally used for

(a) Increasing drug release
(b) Masking taste
(c) Protecting the parent drug from degradation
(d) b & c both

Q 4. Which salt are used maximum in pharma industry

(a) hydrochloride
(b) Sodium
(c) Potassium
(d) Sulphate

Q 5. Solubility can be improved by

(a) Cyclodextrin complexation
(b) Salt formation
(c) Solid dispersion
(d) All of the above.

Q 6. The solubility of the pure form of API is determined by

(a) Partition coefficient
(b) Phase solubility study
(c) Permeability study
(d) All of the above

Q 7. Negative Log of Hydrogen ion concentration is

(a) pOH

(b) pKa

(c) pH

(d) Log P

Q 8. The relationship between pH and the solubility and pKa value of an acidic drug is given by

(a) Henderson Hasselbalch equation

(b) Bragg's Equation

(c) Noye's Whitney Equation

(d) All of the above

Q 9. Most of the drugs are

(a) Strong acids/ base

(b) Neutral

(c) Weak acid/ base

(d) None of the above

Q 10. Strong acids are ionized at

(a) all pH values

(b) pH 1-4 only

(c) pH 7

(d) pH above 10

Answer Key

Q. 1	c	Q. 6	b
Q. 2	d	Q. 7	c
Q. 3	d	Q. 8	a
Q. 4	b	Q. 9	c
Q. 5	d	Q. 10	a

4. Preformulation I (Partition Coefficient & Flow Properties)

Q 1. Lipophilicity of an organic compound is usually described in terms of

(a) Crystallinity
(b) Solubility
(c) pKa
(d) partition coefficient

Q 2. Log P is equal to

(a) [ionized API] org / [unionized API] aq
(b) [unionized API] org / [unionized API] aq
(c) [ionized API] org / [ionized API] aq
(d) [unionized API] org / [ionized API] aq

Q 3. Partition coefficient is determined by the method

(a) Dissolution
(b) Phase solubility
(c) Shake flask method
(d) All of the above

Q 4. Which is Partitioning solvent

(a) Benzene
(b) Ocatanol
(c) Hexane
(d) All of the above

Q 5. Flow property can be improved by

(a) Glidant
(b) Plasticizer
(c) Binder
(d) All of the above.

Q 6. Good flow property is needed for

(a) Easy handling
(b) Good mixing
(c) Content uniformity
(d) All of the above

Q 7. Carr's Index is related to
(a) Flow property
(b) Solubility
(c) Partition coefficient
(d) All of the above

Q 8. This is an indicator of flow property
(a) Carr's Index
(b) Angle of repose
(c) Hausner's ratio
(d) All of the above

Q 9. Flow is excellent, if angle of repose is
(a) 25-30
(b) 36-40
(c) 46-55
(d) None of the above

Q 10. Flow promoters are needed, if angle of repose is
(a) 25-35
(b) 36-40
(c) 46-55
(d) None of the above

Answer Key

Q. 1	d	Q. 6	d
Q. 2	b	Q. 7	a
Q. 3	c	Q. 8	d
Q. 4	d	Q. 9	a
Q. 5	a	Q. 10	c

5. Preformulation II Hydrolysis, Oxidation, Reduction

Q 1. Common chemical degradation reactions are
(a) hydrolysis
(b) oxidation/ reduction
(c) photolysis
(d) All of the above

Q 2. Technique (s) well practiced for protection of drugs from hydrolysis are
(a) Micellization
(b) complexation
(c) Solubilization
(d) All of the above

Q 3. Removal of hydrogen atoms from a carbon atom is
(a) Reduction
(b) Oxidation
(c) Acylation
(d) All of the above

Q 4. Gain of electron is
(a) Reduction
(b) Oxidation
(c) Hydrolysis
(d) All of the above

Q 5. Which is not an antioxidant
(a) Sodium sulphite
(b) Sodium metabisulphite
(c) Sodium sulphate
(d) Sodium thiosulphate

Q 6. Arrhenius equation govern the
(a) Temperature dependence of chemical reactions
(b) Pressure dependence of chemical reactions
(c) Humidity dependence of chemical reactions
(d) All of the above

Q 7. Water vapor carrying capacity of air is called
(a) Partition coefficient
(b) Humidity
(c) Solubility
(d) All of the above

Q 8. Ea, in Arrhenius Equation, is
(a) Activation energy
(b) Electric activity
(c) Activation Electron
(d) None of the above

Q 9. For storage apart from temperature what should be controlled primarily
(a) Light
(b) Humidity
(c) Air
(d) All of the above

Q 10. The drug which is light sensitive
(a) Tetracycline
(b) Indomethacin
(c) Amphotericin B
(d) All of the above

Answer Key

Q. 1	d	Q. 6	a
Q. 2	d	Q. 7	b
Q. 3	b	Q. 8	a
Q. 4	a	Q. 9	b
Q. 5	c	Q. 10	d

6. Preformulation II Racemization

Q 1. Different compounds with same molecular formulae are
(a) Isomers
(b) Cocrystals
(c) Complexes
(d) All of the above

Q 2. Non superimposable mirror images are
(a) Monotropes
(b) Polymorphs
(c) Enantiomers
(d) All of the above

Q 3. Conversion of an optically activesubstance into an optically inactive mixture
(a) Racemization
(b) Oxidation
(c) Reduction
(d) All of the above

Q 4. D and L form of a substance are
(a) Optically in active
(b) Optically active
(c) More bioactive
(d) All of the above

Q 5. Which form of thalidomide is teratogenic
(a) R
(b) S
(c) RS
(d) All of the above

Q 6. In general, which optical isomer is biologically more active
(a) Laevo
(b) Dextro
(c) Racemic
(d) All of the above

Q 7. Optical isomers may differ in the

(a) Taste

(b) Odour

(c) Activity

(d) All of the above

Q 8. Chiral drugs' example

(a) Cisplatin and transplatin

(b) Chloramphenicol palmitate

(c) MDA and Phenacetin

(d) All of the above

Q 9. Molecule that rotate the plane polarized light into the clockwise direction are

(a) Laevo

(b) Dextro

(c) Racemic

(d) All of the above

Q 10. Diastereomers are

(a) Mirror images

(b) Not mirror images

(c) Constitutional isomers

(d) All of the above

Answer Key

Q. 1	a	Q. 6	a
Q. 2	c	Q. 7	d
Q. 3	a	Q. 8	a
Q. 4	b	Q. 9	b
Q. 5	b	Q. 10	b

7. Preformulation II Dissolution, Permeability and BCS

Q 1. Dissolution is

(a) Unofficial test
(b) Official test
(c) Common test
(d) All of the above

Q 2. Dissolution can be expressed by the equation

(a) Modified Noyes-Whitney's
(b) Handerson Hasselbach
(c) Lambert's law
(d) All of the above

Q 3. Dissolution differ from solubility in having

(a) Function of pressure
(b) Function of time
(c) Function of pH
(d) All of the above

Q 4. Dissolution test apparatus type I (IP) is

(a) Basket
(b) Paddle
(c) Paddle over disc
(d) Flow through cell

Q 5. Agitation is given in dissolution test to mimic

(a) Gastric motility
(b) Chewing
(c) Metabolism
(d) All of the above

Q 6. In vitro Permeation study is done by

(a) Dissolution test apparatus
(b) Friabilator
(c) Franz diffusion cell
(d) All of the above

Q 7. Biopharmaceutical Classification System (BCS) is based on
(a) Solubility
(b) Permeability
(c) Dissolution
(d) All of the above

Q 8. Biopharmaceutical Classification System (BCS) classifies drugs into
(a) 3 classes
(b) 4 classes
(c) 2 classes
(d) None of the above

Q 9. Mostly, APIs are from which BCS class
(a) Class I
(b) Class II
(c) Class III
(d) Class IV

Q 10. Highly permeable means, which is absorbed
(a) 85 % or more
(b) 90 % or more
(c) 75 % or more
(d) 95 % or more

Answer Key

Q. 1	b	Q. 6	c
Q. 2	a	Q. 7	d
Q. 3	b	Q. 8	b
Q. 4	b	Q. 9	b
Q. 5	a	Q. 10	a

8. Preformulation II Polymerization

Q 1. Natural polymer is
- (a) HPMC
- (b) PVP
- (c) Guargum
- (d) Ethyl Cellulose

Q 2. Polymer used as binder
- (a) Pthallate
- (b) Azo polymer
- (c) Starch
- (d) All of the above

Q 3. Polymer help in
- (a) Modifying drug release
- (b) binding
- (c) carrier mediated drug delivery
- (d) All of the above

Q 4. PVP stands for
- (a) Poly vinyl pyrrolidone
- (b) Poly vinyl propane
- (c) Piroxy vinyl propane
- (d) Propylene vinyl pyrrolidone

Q 5. HV grade of polymer HPMC is
- (a) High viscosity grade
- (b) Heavy Vinyl grade
- (c) Heavy M. Wt grade
- (d) None of the above

Q 6. Polymers used for enteric coating are
- (a) Sodium Alginate
- (b) Starch
- (c) Cellulose acetate pthallate
- (d) All of the above

Q 7. Guargum may be used for
(a) Colon targeting
(b) Effervescent granules
(c) Fast disintegration
(d) All of the above

Q 8. It is a pH sensitive polymer
(a) Eudragit
(b) PVP
(c) Ethyl Cellulose
(d) HEC

Q 9. Water soluble polymer is
(a) SCMC
(b) EC
(c) Chitosan
(d) All of the above

Q 10. Polymers used in enteric coating release the drugin
(a) acidic pH
(b) Basic pH
(c) Mouth
(d) Gastric region

Answer Key

Q. 1	c	Q. 6	c
Q. 2	c	Q. 7	a
Q. 3	d	Q. 8	a
Q. 4	a	Q. 9	a
Q. 5	a	Q. 10	b

9. Tablet I
Introduction

Q 1. Which are unit dosage form?
(a) Suspension
(b) Tablet
(c) Capsule
(d) b & c both

Q 2. Multi unit dosage form
(a) Tablet
(b) Capsule
(c) Ointment
(d) All of the above

Q 3. One which cannot be delivered by oral tablets
(a) NSAIDs
(b) Antibiotics
(c) Proteins
(d) Harmone

Q 4. Peak and valley effect is associated with
(a) SR dosage forms
(b) Conventional dosage forms
(c) Large volume parenteral products
(d) All of the above

Q 5. Tablets cannot be formulated for API
(a) With very short half life
(b) Which is destroyed in GIT
(c) a & b both
(d) None of the above

Q 6. Nodisintegrant is included in the composition of
(a) Conventional tablets
(b) Chewable tablets
(c) Lozenges
(d) b & c both

Q 7. It can be taken without water
(a) Conventional Tablets
(b) Effervescent Tablets
(c) Chewable Tablets
(d) All of the above

Q 8. Glidants are used for
(a) Improving flow properties
(b) Improving binding
(c) Improved drug release
(d) Improved disintegration

Q 9. The most common granulation method is
(a) Dry granulation
(b) Wet granulation
(c) Direct compression
(d) All of the above

Q 10. Wurster process is associated with
(a) Fluidized Bed Granulation
(b) Dry granulation
(c) Sieving
(d) None of the above

Answer Key

Q. 1	d	Q. 6	d
Q. 2	c	Q. 7	b
Q. 3	c	Q. 8	a
Q. 4	b	Q. 9	b
Q. 5	c	Q. 10	a

10. Tablet II Manufacturing Tablets

Q 1. Tablet punching machine consist of

(a) Hopper
(b) Die and Punch
(c) Feed frame
(d) All of the above

Q 2. In tablet press, granules are charged onto the

(a) Hopper
(b) Feed frame
(c) Punch
(d) All of the above

Q 3. Which is the right flow of action in punching of tablets

(a) Die filling, fill weight management, compaction, ejection
(b) Fill weight management, ejection, Die filling, compaction,
(c) Fill weight management, Die filling, compaction, ejection
(d) Compaction, ejection, Die filling, fill weight management,

Q 4. Tablet manufacturing or the tablet compaction consists of two steps

(a) Compaction and ejection
(b) Granulation, compression
(c) Compression and consolidation
(d) All of the above

Q 5. Partial or complete separation of top or bottom crowns of a tablet is

(a) Lamination
(b) Mottling
(c) Capping
(d) Picking

Q 6. Separation of tablet in two or more distinct layers

(a) Lamination
(b) Mottling
(c) Capping
(d) Picking

Q 7. The tablet defect in which unequal distribution of colors occurs
- (a) Lamination
- (b) Mottling
- (c) Capping
- (d) Double impression

Q 8. Faulty engraving by punches is
- (a) Lamination
- (b) Mottling
- (c) Capping
- (d) Double impression

Q 9. Remedy for Capping is
- (a) Precompression
- (b) Use of suitable colors/dyes
- (c) Mild drying
- (d) All of the above

Q 10. Remedy for mottling is
- (a) Precompression
- (b) Changing solvent system
- (c) Mild drying
- (d) b & c both

Answer Key

Q. 1	d	Q. 6	a
Q. 2	a	Q. 7	b
Q. 3	a	Q. 8	d
Q. 4	c	Q. 9	a
Q. 5	c	Q. 10	d

11. Tablet III

Q 1. Tablet coating is done for

(a) Protection

(b) Masking

(c) Identification

(d) All of the above

Q 2. Which coating increases the weight of tablet significantly

(a) Film Coating

(b) Sugar Coating

(c) Compression coating

(d) All of the above

Q 3. It is a sealing agent

(a) PVP

(b) Shellac,

(c) Starch

(d) Lactose

Q 4. Delayed-release coating (enteric coating)

(a) Only soluble in water at $pH < 5\text{-}6$

(b) Only soluble in water at $pH > 1\text{-}2$

(c) Only soluble in water at $pH \geq 5\text{-}6$

(d) Only soluble in water at $pH \geq 8\text{-}10$

Q 5. Affords flexibility and elasticity to the coat

(a) Polymer

(b) Plasticizer

(c) Solvent

(d) Binder

Q 6. Conventional coating pan is used for

(a) Sugar Coating

(b) Film Coating

(c) Compression coating

(d) a & b both

Q 7. The tablet coating defect which is related to colors
   (a) Blooming
   (b) Mottling
   (c) a & b both
   (d) Orange peel

Q 8. The fading or dulling of a tablet colour is
   (a) Blooming
   (b) Mottling
   (c) a & b both
   (d) Orange peel

Q 9. Compression coating is a
   (a) Dry process
   (b) Wet process
   (c) a & b both
   (d) None of the above

Q 10. Fluidized Bed Dryers is used in
   (a) Granulation
   (b) Drying
   (c) Coating
   (d) All of the above

Answer Key

Q. 1	d	Q. 6	d
Q. 2	b	Q. 7	c
Q. 3	b	Q. 8	a
Q. 4	c	Q. 9	a
Q. 5	b	Q. 10	d

12. Tablet IV QC of Tablets

Q 1. Dissolution Test Apparatus type I USP is

(a) Basket type

(b) Paddle type

(c) Paddle over disc

(d) All of the above

Q 2. Dissolution Test Apparatus type V USP is

(a) Basket type

(b) Paddle type

(c) Paddle over disc

(d) All of the above

Q 3. The capacity of hemispherical jar if dissolution test apparatus is

(a) 900 mL

(b) 500 mL

(c) 1000mL

(d) 100 mL

Q 4. Higuchi plot is the curve of

(a) %CDR V/S Time

(b) Log % Drug remaining V/S Time

(c) % CDR V/S square root of time

(d) Log %CDR V/S Log Time

Q 5. Korsmeyer-Peppasplot is the curve of

(a) %CDR V/S Time

(b) Log % Drug remaining V/S Time

(c) % CDR V/S square root of time

(d) Log %CDR V/S Log Time

Q 6. Weight Variation test is applicable when

(a) the tablet contains 50 mg or more of the drug

(b) the tablet contains 1000 mg or more of the drug

(c) the tablet contains 50 mg or less of the drug

(d) None of the above

Q 7. Content uniformity test is applicable when
(a) the tablet contains 50 mg or more of the drug
(b) the tablet contains 1000 mg or more of the drug
(c) the tablet contains 50 mg or less of the drug
(d) None of the above

Q 8. Disintegration test apparatus contain cylindrical tubes
(a) 2
(b) 4
(c) 6
(d) 8

Q 9. Disintegration time for uncoated tablet is
(a) 15 min
(b) 30 min
(c) 45 min
(d) 60 min

Q 10. In friability test apparatus, tablets are fallen from the height of
(a) 6 inch
(b) 12 inch
(c) 18 inch
(d) 24 inch

Answer Key

Q. 1	a	Q. 6	a
Q. 2	c	Q. 7	c
Q. 3	c	Q. 8	c
Q. 4	c	Q. 9	a
Q. 5	d	Q. 10	a

13. Liquid Orals (Syrups and Elixirs)

Q 1. What is the ideal % of sugar used in syrup-
- (a) 30%
- (b) 65%
- (c) 85%
- (d) None of the above

Q 2. The ideal requirement of pH for oral administration is-
- (a) 4-6
- (b) 5-8
- (c) 6-7
- (d) 8-9

Q 3. Calcium hydroxide solution USP is termed as -
- (a) Lugol's solution
- (b) Lime water
- (c) Lugol water
- (d) None of the above

Q 4. What is the concentration of simple syrup as per BP-
- (a) 55.4% w/w
- (b) 65.3% w/w
- (c) 45.9% w/w
- (d) 67.7% w/w

Q 5. From iodine solution USP is also termed as-
- (a) Lugol's solution
- (b) Lime water
- (c) Both A and B
- (d) Non of the above

Q 6. What is the concentration of simple syrup as per USP-
- (a) 85% w/w
- (b) 45% w/w
- (c) 62% w/w
- (d) None of the above

Q 7. The concentration of sodium benzoate used in syrup preparation is-

(a) 1-2%

(b) 0.6-0.9%

(c) 4-5%

(d) 0.1-0.2%

Q 8. Pharmaceutical Elixirs are -

(a) Aqueous preparation

(b) Non aqueous preparation

(c) Alcoholic preparation

(d) None of the above

Answer Key

Q. 1	c	Q. 5	a
Q. 2	b	Q. 6	a
Q. 3	b	Q. 7	d
Q. 4	d	Q. 8	c

14. Emulsion-I (Introduction Theories and Identification Tests)

Q 1. What is the particle size of the dispersed phase in an emulsion

(a) 1-10 micrometer
(b) 0.1-100 micrometer
(c) 0.01-10 micrometer
(d) 0.01-100 micrometer

Q 2. What are the types of the emulsion

(a) O/W
(b) W/O
(c) O/W/O
(d) All of the above

Q 3. When water is dispersed as globules in oil phase this is termed as-

(a) O/W
(b) W/O
(c) O/W/O
(d) None of the above

Q 4. When oil is dispersed as globules in water phase this is termed as-

(a) O/W
(b) W/O
(c) O/W/O
(d) None of the above

Q 5. Conductivity test is used for identification of which type of emulsion:-

(a) O/W
(b) W/O
(c) Both a and b
(d) None of the above

Q 6. Example of emulsion as a Drug Delivery System are :-

(a) Fluorocarbon
(b) Propofol

(c) Vegetable oil
(d) All of the above

Q 7. Cold cream is example of :-
(a) OW
(b) W/O
(c) O/W/O
(d) All of the above

Q 8. Which type of drugs are quickly released from w/o-
(a) Water soluble
(b) Oil soluble
(c) both a and b
(d) Water insoluble

Q 9. According to viscosity theory increase in viscosity of emulsion leads to:-
(a) Increase in stability
(b) Increase in solubility
(c) Increase in permeability
(d) None of the above

Q 10. Emulsion can be used as-
(a) Orally
(b) Parenterally
(c) Systemically
(d) All of the above

Answer Key

Q. 1	b	Q. 6	d
Q. 2	d	Q. 7	b
Q. 3	b	Q. 8	b
Q. 4	a	Q. 9	a
Q. 5	d	Q. 10	d

15. Emulsion-II (Formulations of Emulsions)

Q 1. How does an emulsifier work?

(a) By being both hydrophilic and hydrophobic

(b) By changing the hydrophobic substance into a hydrophilic substance

(c) By changing the hydrophilic substance into a hydrophobic substance

(d) It equally disperses all materials

Q 2. Which of these is a good example of an emulsifier?

(a) sugar

(b) lecithin

(c) water

(d) oil

Q 3. Increase in the viscosity of continuous phase leads to -

(a) Reduces creamy and coalescence

(b) Increases in creamy and coalescence

(c) All of the above

(d) None of the above

Q 4. The conversion of one type of emulsion to another is termed as-

(a) Phase conversion

(b) sedimentation

(c) Phase inversion

(d) creaming

Q 5. "Larger droplet grow at the expense of smaller one" this phenomenon is termed as:-

(a) Ostwald Ripening

(b) Phase inversion

(c) creaming

(d) none of the above

Q 6. Examples of emulsifying agents are:-

(a) hydrocolloids

(b) Surface active agents

(c) Finally divided solids
(d) All of the above

Q 7. Mechanism of surface active agents is due to reduction in:-
(a) particle size
(b) surface tension
(c) viscosity
(d) All of the above

Q 8. Examples of semi synthetic polysaccharides used as emulsifying agents are-
(a) Methylcellulose
(b) Carboxymethylcellulose
(c) Hydroxypropylmethylcellulose
(d) All of the above

Q 9. Examples of semi synthetic polysaccharides used as emulsifying agents are:-
(a) Carbopol
(b) Polyvinyl alcohol
(c) Polyvinyl pyrolidone
(d) All of the above

Q 10. The hydrophilic lypophilicbalncesystem(HLB) represent no. from -
(a) 1-20
(b) 1-35
(c) 1-18
(d) None of the above

Answer Key

Q. 1	b	Q. 6	d
Q. 2	b	Q. 7	b
Q. 3	a	Q. 8	d
Q. 4	c	Q. 9	d
Q. 5	a	Q. 10	c

16. Suspension Dosage Form

Q 1. Ideal surfactants have HLB value between-

(a) 5-6
(b) 7-8
(c) 7-9
(d) 4-5

Q 2. Suspended particles become flocculated in a suspension, because-

(a) Particles are closely packed
(b) Attractive forces between particles are appreciable
(c) Repulsive forces between particles are appreciable
(d) Vehicles rejects the particles

Q 3. Example of flocculating agents are -

(a) Electrolyte
(b) Surfactants
(c) Polymers
(d) all of the above

Q 4. Example of anionic surfactants are -

(a) Soaps
(b) Alkyl sulfates
(c) Calsoft
(d) All of the above

Q 5. Examples of non-ionic surfactants is/are -

(a) Span
(b) Tergitol
(c) Polyethylene surfactants
(d) All of the above

Q 6. Examples of cationic surfactants are -

(a) Quaternary ammonium salt
(b) Betaines
(c) Both a and b
(d) None of the above

Q 7. Cetrimide and benzalkonium chloride are the examples of-
(a) Non ionic surfactants
(b) Anionic surfactants
(c) Cationic surfactants
(d) All of the above

Q 8. The HLB values of spreading wetting agents is from -
(a) 6-7
(b) 5-8
(c) 7-9
(d) None of the above

Q 9. A suspension can be stabilized by -
(a) By increasing zeta potential
(b) By using surface active agents
(c) By increasing viscosity
(d) All of the above

Q 10. The example/(s) of wetting agent/(s)-
(a) Surfactants
(b) Hydrocolloids
(c) Solvents
(d) All of the above

Answer Key

Q. 1	c	Q. 6	a
Q. 2	b	Q. 7	c
Q. 3	d	Q. 8	c
Q. 4	d	Q. 9	d
Q. 5	d	Q. 10	d

17. Parenterals-I (Introduction: Preformulation of Parenterals)

Q 1. Which of the following dosage form shows maximum bioavailability
- (a) Tablet
- (b) Parenteral
- (c) Emulsion
- (d) None of the above

Q 2. The most important property of parenteral dosage form is
- (a) Sterility
- (b) Clarity
- (c) Permeability
- (d) None of the above

Q 3. The usual volume for intramuscular route of administration is
- (a) 0.5-2ml
- (b) 2-20ml
- (c) 1-4ml
- (d) 5-10ml

Q 4. pKa for acidic drug ranges from --------
- (a) 3-7.5
- (b) 5-8.5
- (c) 7-11
- (d) 11-13

Q 5. X-ray powder diffractrometry of crystalline form gives-------
- (a) Sharp peak
- (b) Diffused peak
- (c) Both sharp/diffused
- (d) None of the above.

Q 6. X-ray powder diffractrometry of amorphous form gives-------
- (a) Sharp peak
- (b) Diffused peak

(c) Both sharp/diffused
(d) None of the above.

Q 7. When several crystals forms of a given chemical exhibits different physical properties. This is called ----------
(a) Polymorphism
(b) Chemical modification
(c) Optical activity
(d) Solvate formation

Q 8. 0.2 – 1 mL is the usual volume, which can be administered by
(a) Intramuscular route
(b) Intravenous route
(c) Intra-cardial route
(d) Intra epidural route

Q 9. Complexation is one of the method to improve ----------
(a) Solubility
(b) Dielectric constant
(c) pKa
(d) pH

Q 10. Which of the physical form gives rapid absorption and slow duration of action
(a) Crystalline form
(b) Amorphous form
(c) Both a and b
(d) None of the above

Answer Key

Q. 1	b	Q. 6	b
Q. 2	a	Q. 7	a
Q. 3	a	Q. 8	c
Q. 4	a	Q. 9	a
Q. 5	a	Q. 10	b

18. Parenterals-I (Formulation of Parenterals)

Q 1. Water for injection (WFI) is used as

(a) Aqueous vehicle

(b) Non-aqueous vehicle

(c) Both a and b

(d) None of the above

Q 2. Polyethylene glycol is generally used as

(a) Preservative

(b) Diluent

(c) Water miscible vehicle

(d) All of the above

Q 3. General storage temperature of water for injection is

(a) 80°c

(b) 40°c

(c) 20°c

(d) 85°c

Q 4. The limit of total organic carbon in water for injection is

(a) 500-600 ppm

(b) 700-1000 ppm

(c) Not more than 500 ppm

(d) None of the above

Q 5. The permissible limit of bacterial endotoxin

(a) More than 0.25 E

(b) Less than 0.25 EU per mm

(c) Both a and b

(d) None of the above.

Q 6. 0.9% of Nacl is used as a standard in pharmaceutical preparation for

(a) Non- pyrogenicity

(b) Sterility

(c) Isotonicity

(d) None of the above

Q 7. Benzalkonium chloride is generally used as

(a) Preservative
(b) Antioxidant
(c) Surfactant
(d) Solubilising agent

Q 8. Which method of preparation for WFI is permitted by European pharmacopeia

(a) Distillation
(b) Reverse osmosis
(c) Ultrafilteration
(d) None of the above

Q 9. What is the value of freezing point (FP) depression of blood and tear?

(a) -0.25°c
(b) 0.25°c
(c) -0.52°c
(d) 0.52°c

Q 10. Which method of adjusting isotonicity is a modified white- Vincent method

(a) Freezing point depression method
(b) Sodium chloride equivalent method
(c) Sprowls method
(d) None of the above

Answer Key

Q. 1	a	Q. 6	c
Q. 2	c	Q. 7	a
Q. 3	a	Q. 8	a
Q. 4	c	Q. 9	c
Q. 5	b	Q. 10	c

19. Parenterals-I (Types) (Types of Parenteral Preparations)

Q 1. Volume of small volume parenteral is-

(a) Less than 100

(b) More than 100

(c) More than 200

(d) Less than 200

Q 2. Volume of Large volume parenteral is-

(a) More than 50

(b) More than 100

(c) More than 1000

(d) All of the above

Q 3. Sterile filtration is generally used for -

(a) Thermostable

(b) Thermoliable

(c) All of the above

(d) None of the above

Q 4. Which parenteral prepration is official in USP-

(a) Sterile ampicillin suspension

(b) Tetanus toxoid

(c) Insulin zinc suspension

(d) All of the above

Q 5. An ideal parenteral suspension should have the following property:-

(a) Syringeability

(b) Injectability

(c) permeability

(d) a and b both

Q 6. Ideal globule size of parenteral emulsion is:-

(a) 0.1-0.5 micrometer

(b) 1-3 micrometer

(c) 1-5 micrometer

(d) 1- 10 micrometer

Q 7. Propofore and oil soluble vitamins are examples of :-

(a) Parenteral suspension

(b) Parenteral emulsion

(c) Parenteral solution

(d) All of the above

Q 8. What is temperature used for Lyophilization-

(a) 0 to -20 °**C**

(b) 0 to 5°**C**

(c) -30 to -40 °**C**

(d) none of the above

Q 9. Parenteral suspension dosage form cannot be administrated by:-

(a) IM Route

(b) IV Route

(c) Ocular route

(d) None of the above

Q 10. The cooling effect in lyophilization is provided by circulating-

(a) Silicone gas

(b) Methane gas

(c) Both a and b

(d) None of the above

Answer Key

Q. 1	a	Q. 6	c
Q. 2	b	Q. 7	b
Q. 3	b	Q. 8	c
Q. 4	d	Q. 9	b
Q. 5	d	Q. 10	a

20. Parenterals-I (Plant Layout) (Plant lay out for Parenterals)

Q 1. What is the full form of HEPA filter-

(a) High efficiency particulate air filter

(b) High efficiency product filter

(c) High efficiency paper filter

(d) None of the above

Q 2. Which method is used for validation of HEPA filter-

(a) DOP test

(b) DOD test

(c) DOC test

(d) None of the above

Q 3. Which area is used for the storage of finally prepared parenteral products-

(a) Preparation area

(b) Clean up area

(c) Quarantine area

(d) None of the above

Q 4. HEPA filter can remove the size of particle up to-

(a) 0.2 micron

(b) 0.5 micron

(c) 1 micron

(d) 0.3 micron

Q 5. What is the storage temperature of water for injection-

(a) Above 120 °C

(b) Below 70 °C

(c) Above 70 °C

(d) None of the above

Q 6. Air control is achieved by installing-

(a) Preliminary air filter

(b) HEPA filters

(c) Both a and b
(d) None of the above

Q 7. The washing of ampoules and vials is carried out in-
(a) Preparation area
(b) Quarantine area
(c) Clean up area
(d) Aseptic area

Q 8. Which Grade area is used to carries out less critical operation in manufacturing of sterile product-
(a) Grade C and A
(b) Grade C and D
(c) Grade A and B
(d) Grade B and D

Q 9. As per the guide lines provide by EU to Good Manufacturing Practices for manufacturing of sterile products. The manufacturing area is classified into-
(a) 5 Grades
(b) 4 Grades
(c) 6 Grades
(d) 2 Grades

Q 10. Water for injection is prepared by-
(a) Distillation
(b) Reverse osmosis
(c) Ultra filteration
(d) All of the above

Answer Key

Q. 1	a	Q. 6	c
Q. 2	a	Q. 7	c
Q. 3	c	Q. 8	b
Q. 4	d	Q. 9	b
Q. 5	c	Q. 10	d

21. Parenterals II (Pyrogens and Pyrogenicity)

Q 1. Standard temp. for autoclaving is -

(a) 120 °C for 1 hr /65 °C for 1 min.

(b) 100 °C for 1 hr /63°C for 1 min.

(c) 140 °C for 5 hr /61°C for 1 min.

(d) 120 °C for 5 hr /66°C for 1 min.

Q 2. LAL test is used for the detection of-

(a) Sterilization

(b) Clarity

(c) Pyrogen

(d) All of the above

Q 3. 1 % methylene blue solution is used for-

(a) Pyrogen test

(b) Leakers test

(c) Sterility test

(d) None of the above

Q 4. Preliminary SHAM test is a part of -

(a) Sterility test

(b) LAL test

(c) Rabbit test

(d) None of the above

Q 5. What is full form of LAL:-

(a) Limulus amebocyte lysate

(b) Leakers Analysis Laboratory test

(c) Limolus polypherous test

(d) None of the above

Q 6. LAL reagent is isolated from :-

(a) Androtoxins

(b) Gram –ve bacteria

(c) Gram +ve bacteria

(d) Horse shoe crab

Q 7. Streptococcus and Clostridium species is the example of:-
(a) Gram +ve bacteria
(b) Gram –ve bacteria
(c) All of the above
(d) None of the above

Q 8. Vibrio and Salmonala species are the examples of :-
(a) Gram +ve bacteria
(b) Gram –ve bacteria
(c) All of the above
(d) None of the above

Q 9. The toxic biological effect of Lipopolysccharide is due to :-
(a) The O-specific chain
(b) The Lipid A
(c) The R-core
(d) None of the above

Q 10. The production of Prostaglandin and Lukotriene are induced by-
(a) antigen
(b) antibody
(c) bacteria
(d) macrophages

Answer Key

Q. 1	a	Q. 6	d
Q. 2	c	Q. 7	a
Q. 3	b	Q. 8	b
Q. 4	c	Q. 9	b
Q. 5	a	Q. 10	d

22. Parenterals II (Sterility Test and Sterilization)

Q 11. The time or dose required for 1 log reduction in microbial population is calculated by-

(a) D value

(b) C value

(c) Z value

(d) F value

Q 12. What is the temperature required in dry heat sterilization for moisture sensitive material-

(a) 160-180°C

(b) 121-134°C

(c) 150-160°C

(d) None of the above

Q 13. What is the temperature required in dry heat sterilization for moisture resistant material-

(a) 160-180°C

(b) 121-134°C

(c) 150-160°C

(d) None of the above

Q 14. Which is most commonlygas used for sterilization-

(a) Carbon dioxide

(b) Nitrous oxide

(c) Ethylene Oxide

(d) Methane

Q 15. Radiation sterilization is carried out by:-

(a) Gamma rays and visual light

(b) Gamma rays and UV light

(c) Beta and Gamma rays

(d) Beta rays and UV light

Q 16. Which method utilizes Hydrogen peroxide as stereilizing method :-

(a) Gas sterilization

(b) Radiation sterilization

(c) Liquid sterilizaiton
(d) Gas and Liquid sterilization

Q 17. The temperature required for 1 log reduction in D value is considered as-
(a) S-value
(b) F-value
(c) D-value
(d) Z-value

Q 18. Which type of material is used for sterilization by Autoclave :-
(a) Thermostable
(b) Thermolabile
(c) All of the above
(d) None of the above

Q 19. What is standard temperature is used for autoclaving :-
(a) 100 °**C** for 1 hr.
(b) 121 °**C** for 1 hr.
(c) 121 °**C** for 30 min.
(d) 100 °**C** for 30 min.

Q 20. Which method is used for sterilization of metal surgical instruments-
(a) Radiation method
(b) chemical
(c) Moist heat
(d) Dry heat

Answer Key

Q. 1	a	Q. 6	c
Q. 2	a	Q. 7	d
Q. 3	b	Q. 8	a
Q. 4	c	Q. 9	b
Q. 5	b	Q. 10	d

23. Capsule I

Q 1. Capsules are?

(a) Unit dosage forms
(b) Solid dosage forms
(c) Compatible with liquid filling material
(d) All of the above

Q 2. Capsule shell is made up of

(a) Gelatin
(b) Carragenan
(c) PVA
(d) All of the above

Q 3. Development of capsule dosage form of an already existing drug in tablet dosage form will be considered as

(a) Strategic reason
(b) Technological reason
(c) Safety reason
(d) Consumer preference

Q 4. The points of interest for the development of a successful capsule dosage form are

(a) Physical and chemical stability
(b) Fill composition
(c) Shell material
(d) All of the above

Q 5. In case of softgel capsules, the gel strength of gelatin should be in the range of

(a) 12-80 bloom
(b) 1500-2000 bloom
(c) 1.5-8 bloom
(d) 150-200 bloom

Q 6. The ratio of drug and plasticizer in case of softgel capsule should be:

(a) 1:5
(b) 1:0.8
(c) 1:0.08
(d) 1:20

Q 7. The right sequence of steps for softgel capsule preparation is:
(a) Drying→Inspection→Encapsulation→Shorting→Cleaning
(b) Encapsulation→Drying→Cleaning→Inspection→Sorting
(c) Cleaning→Drying→Inspection→Encapsulation→Sorting
(d) Cleaning→Drying→Encapsulation →Inspection→Sorting

Q 8. Ribbon system in softgel filling equipment could be of
(a) Pneumatic type
(b) Electronic type
(c) Both
(d) None

Q 9. In-process tests are done
(a) To reduce the cost
(b) To consumer satisfaction
(c) To minimise the risk of product failure
(d) None

Q 10. We cannot use these materials as filling material in softgels
(a) Hygroscopic and volatile material
(b) Aldehydes
(c) Acidic and alkaline
(d) All of the above

Answer Key

Q. 1	d	Q. 6	b
Q. 2	d	Q. 7	b
Q. 3	a	Q. 8	c
Q. 4	d	Q. 9	c
Q. 5	d	Q. 10	d

24. Capsules II

Q 1. A uniform alignment of empty capsule shells with the application of gravity takes place in the process of

(a) Feeding
(b) Rectification
(c) None
(d) All of the above

Q 2. Tamping pin method is associated with

(a) Feeding
(b) Separation
(c) Filling
(d) Sealing

Q 3. Which one is a semiautomatic method of capsule filling:

(a) Auger fill method
(b) Mechanical vibration filling method
(c) Dosator method
(d) Compression filling method

Q 4. A hollow metal tube with a spring loaded adjustable piston is used in

(a) Auger fill method
(b) Mechanical vibration filling method
(c) Dosator method
(d) Compression filling method

Q 5. The proper sequence of the steps involved in sealing method is

(a) Warming →Spraying →Setting
(b) Spraying→Warming→Setting
(c) Warming → Setting → Spraying
(d) Spraying→ Setting → Warming

Q 6. The right combination of active substance and diluent is:

(a) Poorly soluble active, soluble excipient
(b) Soluble active, soluble excipient
(c) Poorly soluble active, insoluble excipient
(d) All of the above

Q 7. The maximum possible amount for capsule filling is:

(a) 450 mg

(b) 500 mg

(c) 600 mg

(d) 650 mg

Q 8. Ribbon system in softgel filling equipment could be of

(a) Pneumatic type

(b) Electronic type

(c) Both

(d) None

Q 9. In-process tests are done

(a) To reduce the cost

(b) To consumer satisfaction

(c) To minimise the risk of product failure

(d) None

Q 10. We cannot use these materials as filling material in hard gelatin capsules

(a) Powdered drug

(b) Granules

(c) Solutions

(d) None of the above

Answer Key

Q. 1	d	Q. 6	b
Q. 2	d	Q. 7	b
Q. 3	a	Q. 8	c
Q. 4	d	Q. 9	c
Q. 5	d	Q. 10	d

25. Capsules III

Q 1. Softgel capsules are most suitable for

(a) Solid filling material

(b) Liquid filling material

(c) Pellet as filling material

(d) All of the above

Q 2. Capsule shell is made up of

(a) Gelatin

(b) Carragenan

(c) PVA

(d) All of the above

Q 3. Development of capsule dosage form of an already existing drug in tablet dosage form will be considered as

(a) Strategic reason

(b) Technological reason

(c) Safety reason

(d) Consumer preference

Q 4. The points of interest for the development of a successful capsule dosage form are

(a) Physical and chemical stability

(b) Fill composition

(c) Shell material

(d) All of the above

Q 5. In case of softgel capsules, the gel strength of gelatin should be in the range of

(a) 12-80 bloom

(b) 1500-2000 bloom

(c) 1.5-8 bloom

(d) 150-200 bloom

Q 6. The ratio of drug and plasticizer in case of softgel capsule should be:

(a) 1:5

(b) 1:0.8

(c) 1:0.08
(d) 1:20

Q 7. The right sequence of steps for softgel capsule preparation is:
(a) Drying→Inspection→Encapsulation→Shorting→Cleaning
(b) Encapsulation→Drying→Cleaning→Inspection→Sorting
(c) Cleaning→Drying→Inspection→Encapsulation→Sorting
(d) Cleaning→Drying→Encapsulation →Inspection→Sorting

Q 8. Ribbon system in softgel filling equipment could be of
(a) Pneumatic type
(b) Electronic type
(c) Both
(d) None

Q 9. In-process tests are done
(a) To reduce the cost
(b) To consumer satisfaction
(c) To minimise the risk of product failure
(d) None

Q 10. We cannot use these materials as filling material in softgels
(a) Hygroscopic and volatile material
(b) Aldehydes
(c) Acidic and alkaline
(d) All of the above

Answer Key

Q. 1	d	Q. 6	b
Q. 2	d	Q. 7	b
Q. 3	a	Q. 8	c
Q. 4	d	Q. 9	c
Q. 5	d	Q. 10	d

26. Capsules IV

Q 1. What should be the moisture content of a hard gelatin capsule:

(a) <10%

(b) 10-13%

(c) 13-16%

(d) >16%

Q 2. What should be the moisture content of a soft gelatin capsule:

(a) <5%

(b) 6-10%

(c) 8-10%

(d) >10%

Q 3. FTIR, assay and limit tests are mandatory for the testing of:

(a) Finished product

(b) Raw material

(c) In process material

(d) None of the above

Q 4. Bloom strength is a prime test for

(a) Powdered gelatin

(b) Empty capsule shell

(c) Filled capsule shell

(d) Gelatin solution

Q 5. Test for antimicrobial presence is mandatory for

(a) Gelatin

(b) Hypromellose

(c) Both

(d) None

Q 6. Bloom strength is tested using the instrument called:

(a) Rheometer

(b) Gleometer

(c) Bloom meter

(d) All

Q 7. What should be the range of pH for Hypromellose capsules:

(a) 3.2-4.8
(b) 4.8-5.8
(c) 5.8-8.0
(d) 8.0-9.2

Q 8. Ribbon system in softgel filling equipment could be of

(a) Pneumatic type
(b) Electronic type
(c) Both
(d) None

Q 9. Which type of dissolution instrument is used for capsule dosage forms:

(a) Paddle type
(b) Basket type
(c) Both
(d) None

Q 10. Which is a suitable type packaging for capsules:

(a) Strip
(b) Blister
(c) Both
(d) None

Answer Key

Q. 1	c	Q. 6	b
Q. 2	b	Q. 7	c
Q. 3	b	Q. 8	c
Q. 4	d	Q. 9	b
Q. 5	a	Q. 10	c

27. Pellets & Pelletization

Q 1. Introduction of seeds or nuclei is the first step of __________.

(a) Cryopelletization
(b) Solution/suspension layering
(c) Powder layering
(d) Extrusion–spheronization

Q 2. Air distribution system to cause fluidization is a part of __________.

(a) Cryopelletization
(b) Solution/suspension layering
(c) Powder layering
(d) Extrusion–spheronization

Q 3. Counter rotating double cylinders are a part of __________.

(a) Cryopelletization
(b) Solution/suspension layering
(c) Powder layering
(d) Extrusion–spheronization

Q 4. Ultra-low temperature to formulate solid spheres takes place in____.

(a) Cryopelletization
(b) Solution/suspension layering
(c) Powder layering
(d) Extrusion–spheronization

Q 5. Which one of these is not a pelletization process?

(a) Spray drying/congeling
(b) Melt spheronization
(c) Fluidised granulation
(d) Extrusion–spheronization

Answer Key

Q. 1	c	Q. 4	a
Q. 2	b	Q. 5	c
Q. 3	d		

28. Ophthalmic Preparations Introduction, Absorption through Eye, Formulation Considerations

Q 1. Ophthalmic dosage forms instilled onto the external surface of the eye are called?

(a) Intraocular
(b) Periocular
(c) Topical
(d) None of the above

Q 2. Ophthalmic dosage forms administered inside the eye by injection are called?

(a) Intraocular
(b) Periocular
(c) Topical
(d) None of the above

Q 3. Ophthalmic dosage forms administered adjacent to the eye are called?

(a) Intraocular
(b) Periocular
(c) Topical
(d) None of the above

Q 4. The _______ chamber has gel-like gelatinous fluid.

(a) Aqueous
(b) Vitreous
(c) Both
(d) None of the above

Q 5. When the drug is dispersed in the solvent and exists in fine particulate form, the formulation is called

(a) Ointment
(b) Suspension
(c) Powder
(d) Granules

Q 6. What factor does not affect ophthalmic drug availability?

(a) Gravity

(b) Lacrimation

(c) Blinking

(d) food

Answer Key

Q. 1	c	Q. 4	b
Q. 2	a	Q. 5	b
Q. 3	b	Q. 6	d

29. Ophthalmic Preparations Dosage Form

Q 1. In the preparation of eye drops involves multiple steps. The correct sequence of these steps is?

(a) Preparation→Clarification→Filling→Sterilization
(b) Preparation→Filling→Clarification→Sterilization
(c) Preparation→Sterilization→Clarification→Filling
(d) Sterilization→Preparation→Clarification→Filling

Q 2. Concentration of dissolved drug cannot be manipulated due to their ________?

(a) Relative solubility
(b) Relative insolubility
(c) Relative viscosity
(d) All the above

Q 3. Which one of these is not a requirements of eye lotions?

(a) Sterile
(b) Preservative-free
(c) Isotonic
(d) Acidic

Q 4. Which one of these is not an advantage of ocular inserts?

(a) Increasing contact time
(b) Preservative property
(c) Better efficacy
(d) Better patient compliance

Q 5. ______ is a part of the eye?

(a) Anterior chamber
(b) Posterior chamber
(c) Vitreous body
(d) All

Answer Key

Q. 1	a	Q. 4	b
Q. 2	b	Q. 5	d
Q. 3	d		

30. Ophthalmic Preparations Dosage Form

Q 1. What is a newer dosage forms for ophthalmic drug delivery?
- (a) Solutions
- (b) Suspensions
- (c) Intravitreal injections
- (d) Ointments

Q 2. What acts as a lipid barrier?
- (a) Lachrymation
- (b) Low corneal permeability
- (c) Blinking reflex
- (d) Peripheral blood flow

Q 3. Preservatives are included in unit-dose package.
- (a) True
- (b) False
- (c) Not sure
- (d) Don't know

Q 4. UV transmittance is a method of testing for?
- (a) Tonicity
- (b) Viscosity
- (c) Clarity
- (d) Sterility

Q 5. Which one of these is not a part of Franz diffusion cell?
- (a) Donor compartment
- (b) Membrane filter
- (c) Semipermeable membrane
- (d) Receiver compartment

Answer Key

Q. 1	c	Q. 4	c
Q. 2	c	Q. 5	b
Q. 3	b		

31. Pharmaceutical Aerosols I

Q 1. Common household example of propellant is/are?

(a) Deospray

(b) Shaving foam

(c) Pain killer spray

(d) All of the above

Q 2. The most critical component of MDIs

(a) Canister

(b) Metering valve

(c) Expansion Chamber

(d) None of the above

Q 3. This is not the advantage of aerosol

(a) Cheap

(b) Portable

(c) Direct delivery to site of action

(d) Quick onset of action

Q 4. In aerosols drug is …….. in the propellant

(a) Dissolved

(b) Emulsified

(c) Suspended

(d) All of the above

Q 5. Aerosol provides

(a) Local effect only

(b) Systemic effect only

(c) Local and Systemic Effect both

(d) All of the above

Q 6. Advantage/(s) of Aerosol

(a) Dose to dose maintenance of sterility

(b) Minimum dose required

(c) Quick onset of action

(d) All of the above

Q 7. Drugs which are used in aerosol.
(a) Nitroglycerin
(b) Hydrocortisone
(c) Cromolyn Sodium
(d) All of the above

Q 8. Oral aerosols delivery may be done for the drug/(s)
(a) Methyl salicylate
(b) Nitroglycerin
(c) Nonoxyenol-9
(d) All of the above

Q 9. Example of drugs used in Respiratory Aerosols
(a) Methyl salicylate
(b) Albuterol
(c) Nonoxyenol-9
(d) All of the above

Q 10. Example of drugs used in Nasal Aerosols
(a) Methyl salicylate
(b) Nonoxyenol-9
(c) Cromolyn Sodium
(d) All of the above

Answer Key

Q. 1	d	Q. 6	d
Q. 2	b	Q. 7	d
Q. 3	a	Q. 8	b
Q. 4	d	Q. 9	b
Q. 5	c	Q. 10	c

32. Pharmaceutical Aerosols II (Components of Aerosols)

Q 1. Which is not a component of aerosol?
(a) Propellant
(b) Valve and actuator
(c) Hopper
(d) Container

Q 2. Heart of an aerosol package is
(a) Canister
(b) Metering valve
(c) Propellant
(d) None of the above

Q 3. This is not the liquefied propellant
(a) Hydrocabons
(b) Hydrofluoroalkanes
(c) Chloroflurocarbons
(d) Compressed Gases

Q 4. Propellants which deplete ozone
(a) Hydrocabons
(b) Hydrofluoroalkanes
(c) Chloroflurocarbons
(d) Compressed Gases

Q 5. Final pressure is equal to initial pressure with propellants
(a) CFC11
(b) Propane
(c) Carbon di oxide
(d) a & b both

Q 6. Dichlorodifluoromethaneis
(a) CFC114
(b) CFC12
(c) CFC11
(d) HFA227

Q 7. CFCs are replaced by
  (a) HFAs
  (b) HCs
  (c) Compressed gases
  (d) All of the above

Q 8. Aerosol made up of coated glass can withstand pressureupto
  (a) 18 psig
  (b) 25 psig
  (c) 140 psig
  (d) 180 psig

Q 9. Pressure drops with the usage with aerosols containing propellants
  (a) Hydrocabons
  (b) Hydrofluoroalkanes
  (c) Chloroflurocarbons
  (d) Compressed Gases

Q 10. First digit in the digital name of liquefied propellant refers to
  (a) Number of carbon − 1
  (b) Number of hydrogen − 1
  (c) Number of carbon + 1
  (d) Number of hydrogen +1

Answer Key

Q. 1	c	Q. 6	b
Q. 2	c	Q. 7	a
Q. 3	d	Q. 8	b
Q. 4	c	Q. 9	d
Q. 5	d	Q. 10	a

33. Pharmaceutical Aerosols III (Components and Systems of Aerosols)

Q 1. Dip tube is the part of
(a) Propellant
(b) Valve
(c) Actuator
(d) Container

Q 2. It releases the content of aerosol
(a) Canister
(b) Metering valve
(c) Propellant
(d) Actuator

Q 3. Higher doses can be delivered by
(a) Suspension systems
(b) Solution systems
(c) Emulsion systems
(d) None of the above

Q 4. The stability of aerosol dispersions is improved by
(a) Lower the Moisture content (< 300 ppm)
(b) Reduce particle size < 5 μ (50-100 μ for topical aerosols)
(c) a & b both
(d) Surfactant with HLB < 10

Q 5. High initial pressure is needed with propellants
(a) Compressed gas systems
(b) Liquefied gas systems
(c) Barrier type system
(d) a& b both

Q 6. Propellants are emulsified in these systems
(a) Two phase systems
(b) Barrier type systems
(c) Foam systems
(d) None of the above

Q 7. Aerosol is manufactured by
   (a) Pressure filling
   (b) Cold filling
   (c) a & b both
   (d) None of the above

Q 8. Three Phase system is also called
   (a) Solution system
   (b) Water based system
   (c) Suspension system
   (d) All of the above

Q 9. Which component/part govern the characteristics of dispensed product
   (a) Propellant
   (b) Valve
   (c) Actuator
   (d) Container

Q 10. Inner diameter of dip tube may be
   (a) upto 0.195 inch
   (b) more than 0.195 inch
   (c) Less than 0.120 inch
   (d) None of the above

Answer Key

Q. 1	b	Q. 6	c
Q. 2	d	Q. 7	c
Q. 3	a	Q. 8	b
Q. 4	c	Q. 9	b
Q. 5	a	Q. 10	a

34. Pharmaceutical Aerosols IV (Inhalers and Evaluation of Aerosols)

Q 1. Which type of inhaler is propellant free?
(a) MDI
(b) SVN
(c) DPI
(d) None of the above

Q 2. The most critical component of MDIs
(a) Canister
(b) Metering valve
(c) Expansion Chamber
(d) None of the above

Q 3. The piezoelectric effect is associated with
(a) MDI
(b) Pneumatic jet nebulizer
(c) Mesh Nebulizer
(d) Ultrasonic nebulizer

Q 4. Plate with multiple apertures to produce a liquid aerosol is used in
(a) MDI
(b) Pneumatic jet nebulizer
(c) Mesh Nebulizer
(d) Ultrasonic nebulizer

Q 5. Tag Open Cap Apparatus is used in
(a) Flame projection test
(b) Flash point
(c) Delivery rate
(d) Leakage test

Q 6. Particle size distribution in aerosols is evaluated by
(a) SEM
(b) Sieving
(c) Cascade Impactor
(d) None of the above

Q 7. Flame Projection test involves spraying on flame for .... Sec.
(a) 2
(b) 4
(c) 8
(d) 15

Q 8. DPI stands for
(a) Dry Pressure Inhaler
(b) Drug Pressure Inhaler
(c) Dry Powder Inhaler
(d) Drug Powder Inhaler

Q 9. Fixed volume range in MDI is
(a) 25- 50 mL
(b) 25- 150 microL
(c) 50-100 mL
(d) 50-100 microL

Q 10. Vapor ignition temperature is noted as
(a) Flame projection
(b) Flash point
(c) Delivery rate
(d) Minimum fill

Answer Key

Q. 1	c	Q. 6	c
Q. 2	b	Q. 7	b
Q. 3	d	Q. 8	c
Q. 4	c	Q. 9	b
Q. 5	b	Q. 10	b

35. Cosmetics-I

Q 1. The skin of an average body covers a surface area of approximately
- (a) 2.0 sq.m.
- (b) 1.5sq.m.
- (c) 2.2sq.m.
- (d) 1.0 sq.m.

Q 2. Cold cream is
- (a) W/O emulsion
- (b) O/W emulsion
- (c) Both a and b
- (d) None of the above

Q 3. The ideal pH value of cold cream is
- (a) 2.4 - 5
- (b) 5.6 – 8
- (c) 4.6 - 6
- (d) 3.6 - 7

Q 4. Which is the special ingredient bused to remove finelines in cream
- (a) Lecithin
- (b) Lauryl sulphate
- (c) Lactic acid
- (d) Glycerine

Q 5. Hydroquinones are used as a
- (a) Preservatives
- (b) Buffer
- (c) Bleaching agent
- (d) None of the above

Q 6. Vanishing cream is
- (a) O/W emulsion
- (b) W/O emulsion
- (c) All of the above
- (d) None of the above

Q 7. Which is the major ingredient is used in the preparation of vanishing cream

(a) Polypropylene
(b) Stearic acid
(c) Isopropyl stearate
(d) Mineral oil

Q 8. Which is the commonly used humectant in cosmetic preparation

(a) Polyethylene glycol
(b) IPA (Isopropyl alcohol)
(c) Glycerol
(d) None of the above

Q 9. Which type of cream gets disappear on rubbing over skin

(a) Vanishing cream
(b) Cold cream
(c) Both a and b
(d) None of the above

Q 10. Which is the commonly used emulsifier in cream preparation

(a) Borax
(b) Alcohol
(c) Glycerine
(d) Lanoline

Answer Key

Q. 1	a	Q. 6	a
Q. 2	a	Q. 7	b
Q. 3	b	Q. 8	c
Q. 4	c	Q. 9	a
Q. 5	c	Q. 10	a

36. Cosmetics-II (Sunscreen Preparations and Dentifrices)

Q 1. SPF stands for
(a) Sun Permeation Factor
(b) Sun Product factor
(c) Sun Perfection factor
(d) Sun Protection factor

Q 2. The ideal value of SPF for a sunscreen preparation is
(a) Atleast 20
(b) Atleast 15
(c) Atleast 17
(d) None of the above

Q 3. What is the ideal percentage of detergent used in dentifrices
(a) 1 - 5%
(b) 1 – 9 %
(c) 1 – 3 %
(d) None of the above

Q 4. What is the ideal percentage of humectant used in dentifrices
(a) 20 – 35%
(b) 20 - %45
(c) 10 – 35%
(d) 10 – 25%

Q 5. What is the ideal percentage of sweetening agent used in dentifrices
(a) 0.09 – 3%
(b) 2 - 3%
(c) 1 – 2%
(d) 0.05 – 2%

Q 6. Sodium monofluorophosphate (Na2PO3F) is used as --------------
(a) Anti-tartar agent
(b) Anti-cavity agent
(c) Anti-plaque
(d) None of the above

Q 7. Hydroquinones are used as a
- (a) Preservatives
- (b) Buffer
- (c) Bleaching agent
- (d) None of the above

Q 8. The formula for the calculation of SPF is
- (a) SPF= MED of photoprotected skin / MED unprotected skin
- (b) SPF= MED of unprotected skin / MED of photoprotected skin
- (c) Both a and b
- (d) None of the above

Q 9. Sunscreens preparations filter out UV rays in the region of -----------
- (a) 2900 – 3300 Angstroms
- (b) 2500 – 3000 Angstroms
- (c) 3900 – 4000 Angstroms
- (d) All of the above

Q 10. Paraminobenzoic acid (PABA) is one of the examples of
- (a) Physical sunscreen agent
- (b) Organic sunscreen agent
- (c) Both a and b
- (d) None of the above

Answer Key

Q. 1	d	Q. 6	b
Q. 2	b	Q. 7	c
Q. 3	c	Q. 8	a
Q. 4	a	Q. 9	a
Q. 5	a	Q. 10	b

37. Cosmetics III Shampoo, Hair Dye and Lipstick

Q 1. The creamy appearance of shampoo is due to-

(a) Pearlizers and opacifiers

(b) Conditioning agent

(c) Aesthetic additives

(d) None of the above

Q 2. Glutamic acid derivatives are examples of-

(a) Aesthetic

(b) Conditioning agent

(c) Thickening agent

(d) Cleansing agent

Q 3. The most commonly cleansing agent used in the preparation of shampoo is-

(a) Detergents

(b) Silicon additives

(c) Amino acid

(d) None of the above

Q 4. The acceptable limit for ph of a shampoo is-

(a) 5-6

(b) 8-9

(c) 2-3

(d) 4-9

Q 5. Example of pigments and dies used in the preparation of lipstick is-

(a) Manganese violet

(b) Titanium Di-oxide

(c) D and C red no. 6

(d) All of the above

Q 6. The ideal property of temporary hair colour is-

(a) Do not penetrate into the hair

(b) Can be easily rinsed off

(c) Both a and b
(d) None of the above

Q 7. Shampoo's provide cleansing action to hair by removing-
(a) Dust
(b) Oil
(c) Sebum
(d) All of the above

Q 8. Cellulose derivate's are used in the preparation of shampoo as-
(a) Thickening agent
(b) Pearlizers
(c) Cleansing agent
(d) Non of the above

Q 9. Stalagmometer is used to determine-
(a) viscosity
(b) Surface tension
(c) Both a and b
(d) Non of the above

Q 10. Sachets packing of shampoo is used because it-
(a) Save money
(b) Easy to carry
(c) flexible
(d) All of the above

Answer Key

Q. 1	a	Q. 6	c
Q. 2	a	Q. 7	d
Q. 3	a	Q. 8	a
Q. 4	d	Q. 9	b
Q. 5	d	Q. 10	d

38. Packaging Materials Science- I Materials

Q 1. Packaging is the?
(a) Art
(b) Technology
(c) Science
(d) All of the above

Q 2. Which is not the function of packaging
(a) Protection
(b) Masking
(c) Identification
(d) Appearance

Q 3. Packaging should help in
(a) presentation
(b) Information
(c) Identification
(d) All of the above

Q 4. Primary package is
(a) Vial
(b) Ampoule
(c) Blister pack
(d) All of the above

Q 5. Secondary package is
(a) Cartons
(b) Paper box
(c) Blister pack
(d) Ampoule

Q 6. Tertiary package is
(a) Cartons
(b) Paper box
(c) Blister pack
(d) Ampoule

Q 7. Package information is governed by
(a) Pharmacy Act
(b) Drugs and Cosmetic Act
(c) Dangerous drugs Act
(d) All of the above

Q 8. A container that is impervious to air or any other gas under normal conditions
(a) Hermetically Sealed
(b) Airtight container
(c) Tamper-evident container
(d) All of the above

Q 9. Highly resistant glass is
(a) Type I
(b) Type II
(c) Type III
(d) NP

Q 10. Highly resistant glass contain
(a) High % silicate
(b) High % boron
(c) High % aluminium
(d) High % sodium

Answer Key

Q. 1	d	Q. 6	a
Q. 2	b	Q. 7	b
Q. 3	d	Q. 8	a
Q. 4	d	Q. 9	a
Q. 5	b	Q. 10	b

39. Packaging Materials Science- II Official Requirements & Stability Aspects

Q 1. What is the cost effective way of product innovation?

(a) New API development

(b) New package development

(c) a & b both

(d) None of the above

Q 2. Example of incorporating a device into the pack

(a) Insulin syringe

(b) MDI

(c) Microsyringe

(d) All of the above

Q 3. Tamper-evident packaging is/are

(a) Blister pack

(b) Heat sealed tubes

(c) a & b both

(d) None of the above

Q 4. Child-resistant closures works as

(a) The "press–turn"

(b) The "squeeze–turn"

(c) A combination lock

(d) All of the above

Q 5. Which is not child resistant technology

(a) Push and turn down.

(b) Squeeze the sides and turn.

(c) Push down the tab and turn

(d) Turn clockwise

Q 6. Which is the Tamper-evident technology

(a) Heat-shrunk bands or wrappers

(b) Aerosol container

(c) Bottles with inner mouth seals

(d) All of the above

Q 7. Child-resistant closures should be

(a) difficult to be opened by young children

(b) difficult to be opened by adults

(c) easy to be opened by adults

(d) a &c both

Q 8. Which is not the stability test of packages

(a) Compression Strength Testing

(b) Distribution Simulation Testing

(c) Package Integrity Testing

(d) Discharge Rate Testing

Q 9. Package Integrity Testing is performed for

(a) dye leak,

(b) visual inspection,

(c) vacuum leak

(d) All of the above

Q 10. Package strength testing involves

(a) visual inspection,

(b) vacuum leak

(c) burst testing

(d) a & b both

Answer Key

Q. 1	b	Q. 6	d
Q. 2	b	Q. 7	d
Q. 3	c	Q. 8	d
Q. 4	d	Q. 9	d
Q. 5	d	Q. 10	c

40. Packaging Materials Science- III QC Tests of Packaging Materials

Q 1. Which is not the glass test?

(a) Water attack test.

(b) Water vapor permeability

(c) Arsenic limit test.

(d) Light transmission test

Q 2. Alkali leaching from glass is assessed by

(a) Water attack test.

(b) Water vapor permeability

(c) Arsenic limit test.

(d) Light transmission test

Q 3. Water Attack Test USP is similar with the IP test

(a) Hydrolytic resistance Test

(b) Water vapor permeability

(c) Arsenic limit test.

(d) None of the above

Q 4. Indicator used in hydrolytic resistance test IP is

(a) Phenolphthalein

(b) Methyl red

(c) Thymol blue

(d) All of the above

Q 5. Volume of acid used in Powdered Glass Test should be least for

(a) Type I glass.

(b) Type II glass

(c) Type III glass

(d) Type NP glass

Q 6. The common QC test for non-injectable and injectable Plastic Containers

(a) Collapsibility test

(b) Reducing substances

(c) Light absorption
(d) Water vapor permeability

Q 7. Self sealability is the QC test for
(a) Glassclosures
(b) Plasticclosures
(c) Rubber closures
(d) a &c both

Q 8. Maximum number of fragments allowed in Fragmentation test
(a) 15
(b) 5
(c) 10
(d) 20

Q 9. In pH of aqueous extract test the volume limit of NaOH is
(a) Not more than 0.9 mL
(b) Not more than 0.6 mL
(c) Not more than 0.3 mL
(d) Not more than 0.8 mL

Q 10. Which acid is used in Water Attack Test
(a) Hydrochloric Acid
(b) Sulphuric Acid
(c) Nitric Acid
(d) a & b both

Answer Key

Q. 1	b	Q. 6	a
Q. 2	a	Q. 7	c
Q. 3	a	Q. 8	a
Q. 4	b	Q. 9	c
Q. 5	a	Q. 10	b

www.ingramcontent.com/pod-product-compliance
Ingram Content Group UK Ltd.
Pitfield, Milton Keynes, MK11 3LW, UK
UKHW021531300726
14060UKWH00011B/360

9 789391 910327